Sirius

Brightest Diamond in the Night Sky

Jay B. Holberg

Sirius

Brightest Diamond in the Night Sky

Dr Jay B. Holberg
Senior Research Scientist
Lunar and Planetary Laboratory
University of Arizona
Tucson
Arizona
U.S.A.

SPRINGER–PRAXIS BOOKS IN POPULAR ASTRONOMY
SUBJECT *ADVISORY EDITOR*: John Mason B.Sc., M.Sc., Ph.D.

ISBN 10: 0-387-48941-X Springer Berlin Heidelberg New York
ISBN 13: 978-0-387-48941-4 Springer Berlin Heidelberg New York

Springer is part of Springer-Science + Business Media (springer.com)

Library of Congress Control Number: 2006938990

Printed in Germany

Cover design: Jim Wilkie
Project management: Originator Publishing Services, Gt Yarmouth, Norfolk, UK

Printed on acid-free paper

Contents

Acknowledgments

The idea for a book on the star Sirius occurred to me in April 2001 while I was organizing some old 19th century observations of the orbit of the star's companion. At the time I had only a few vague notions of just how rich and varied this subject would become. As coincidence would have it, I was not alone in my interests in Sirius. Dr. Hugh Van Horn (National Science Foundation, retired), an expert on the stellar interiors and someone who has trained numerous graduate students in the study of white dwarfs, was also making similar plans. We became aware of each other's efforts at a White Dwarf Workshop in Naples, Italy in the summer of 2002. I remain most grateful to Hugh for generously allowing me to proceed, for his encouragement of my efforts, and for his careful reading of an early draft of this book. I am also very indebted to Professor François Wesemael of the University of Montreal, for his interest and encouragement in this project, for his comments on early drafts of this work, and for his help with French translations, including the Chacornac logbook from the Observatoire de Paris.

Many others have provided valuable help with various aspects of this book. Professor Volker Weidemann, Kiel University, was instrumental in helping to locate several sources of material from the early days of white dwarf research. Dr. Lotfi Ben Joffel of the Institut d'Astrophysique de Paris-CNRS (IAP) assisted with Arabic and Koranic material on Sirius, as did Bassel Reyahi, who provided a copy of an English summary of his book *Sirius: A Scientific and Qur'anic Perspective*. Dr. Alain Lecavelier of the Institut d'Astrophysique de Paris-CNRS (IAP) provided a very enjoyable tour of the Observatoire de Paris. Dr. Ben Joffel also helped with arrangements for my visit to the Observatoire de Paris library.

Much of the effort in writing this book involved my immersion into several subject areas in which I have had no professional training; this is particularly true of the Dogon material. I have profited enormously from discussions with John

McKinney, of Bamako, Mali. His intimate knowledge of the Dogon and the history of missionary work in Mali in the early 20th century was of great value. Dr. Geneviève Calame-Griaule, the daughter of Marcel Griaule, contributed her valuable insights into her father's work among the Dogon. Dr. Walter van Beek of the African Studies Centre, in Leiden, also shared his views and comments on the Dogon controversy. Silvio Bedinni and Erich Martel were very helpful with material and background on Benjamin Bannekar. Dr. Richard Wilkinson of the Classics Department of the University of Arizona provided his expert advice on the world of the Egyptians, through a careful reading of Chapter 1. Dr. Michael R. Molnar (Rutgers University, retired) provided his unique insights on the coinage of the ancient world, as well as superb images of ancient coins from his collection. Ms. Kate Magargal helped with some of the research for this book, and also provided material from her independent study project at the University of Arizona on the stellar lore and navigation techniques of the Polynesians. Dr. Jean-François Mayer, of the Freiberg University in Switzerland, supplied insight and suggestions on the subject of the Solar Temple.

A great deal of the material in this book comes from hard-to-find publications and 19th century archives. In this regard I wish to thank Brenda Corbin and Gregory Shelton at the U.S. Naval Observatory in Washington, D.C., for helping me locate many very difficult references as well as Ms. Mary Guierri at the National Optical Astronomy Observatories Library in Tucson for her patience and assistance. Some of the material on the discovery of the companion of Sirius, involving G. P. Bond and Alvan Clark, comes from the Harvard College Observatory Archives and I wish to thank Ms. Alison Doane, for her assistance with Harvard Plate files and the observatory logs. I am also most grateful to Josette Alexandre of the Library at the Observatoire de Paris for locating and providing copies of records from Jean Chacornac's 1862 observing logs and for facilitation of my visit to the Library. Thanks are also due Dr. Dan Lewis of the Huntington Library in San Marino, California, for providing copies of the correspondence between Walter Adams, Arthur Eddington, and Ejnar Hertzsprung. Additional thanks is due to Dr. Alan Batten of the Herzberg Institute of Astrophysics in Victoria, BC, for locating the image of Otto Struve and to Richard Dreiser of Yerkes Observatory and Alison Doane for providing other historical images.

A number of people were of great help in reading, correcting, and commenting on early drafts of this book. I particularly wish to thank my wife, Dr. Catharine J. Holberg, for her constant encouragement and patient reading of all of my drafts. Her questions and alternative suggestions of my explanations of various topics helped me enormously in shifting from my habit of writing for other astronomers, to that of the general public. I also wish to thank fellow physicist Dr. Bradford Barber, of the Nuclear Medicine Department of the University of Arizona, whose insightful comments and encouragement were of immense help. Professor James Liebert of Steward Observatory, the University of Arizona, also contributed many insightful comments on my final chapter. Dr. Bradley Schaefer of the Louisiana State University suggested a number of corrections and changes to Chapter 2. I wish to express my gratitude to Dr. Simon Mitton of St. Johns College, Cambridge, for his encouragement and wise council in helping to get this book published. Finally, much gratitude is due Clive Horwood of Praxis Publishing for the expeditious and professional manner in which this book was brought to press.

Introduction

This book was written to tell two stories. The first and most obvious is why the star known as Sirius has been regarded as an important fixture of the night sky by many civilizations and cultures, since the beginnings of history. A second, but related, narrative is the prominent part that Sirius has played in how we came to achieve our current scientific understanding of the nature and fate of the stars. These two topics have a long intertwined history, and the telling of one story eventually leads back to the other. Presently, new observations from space are revealing, in precise terms, how stars like Sirius and the sun have evolved and what they will ultimately become, while at the same time answering some of the age-old questions about Sirius.

The introduction to this book is a particularly good place to establish some of the well-known facts regarding Sirius. Of all the fixed stars in the night sky, Sirius is by far the brightest—almost twice as bright as its nearest rival, the star Canopus, which resides too far south to be viewed over much of the northern hemisphere. Only the sun, the moon, and at times the planets Venus, Jupiter, and on rare occasions Mars, appear brighter. Sirius, with its flashing brilliance, is a striking feature of the northern winter sky and has understandably drawn the attention of observers of the heavens for thousands of years. Sirius can be easily seen over most of the surface of the earth, except for a zone north of the Arctic Circle, where it never rises. Every year Sirius emerges from the glare of the sun in late July, and by year's end it hangs on the meridian at midnight. In late May it disappears from view, rejoining the sun for a period of slightly over two months. It was around this annual coming and going that certain ancient cultures and civilizations organized their religious year, and synchronized their calendars with the agricultural cycles. It was also surrounding Sirius that a host of elaborate legends and beliefs came into being.

Sirius has many names, astronomers recognize over fifty designations for the star, but the most prominent is Alpha Canis Majoris, the brightest star in the constellation Canis Major, Latin for the great dog. Over the centuries many beliefs have come to be associated with Sirius. Some of these beliefs still echo in such phrases as "the dog days of summer", which the ancient Romans understood well. Other old beliefs long ago fell from public consciousness—only to be revived and to grow into modern popular and scientific controversies. Still other notions regarding Sirius abound in various forms in the occult and new-age sections of bookstores—and in even greater profusion on the Internet. Although these beliefs may seem quite recent, many have their origins in the ancient lore surrounding Sirius: humans seem naturally drawn to its brilliance, and a surprising number of modern cults have nucleated around beliefs in which Sirius plays a prominent role.

Today we know Sirius as a rather typical star, with a surface temperature nearly twice that of the sun. The average temperature of the visible outer layers of its atmosphere is approximately 9400 K (16,900 F): this explains its striking blue–white appearance. Its apparent brightness is due to its relative proximity to the earth of about 8.6 light years. There are relatively few stars as hot as Sirius in the solar neighborhood. The most similar nearby star, Vega, lies at a distance of 26 light years from the sun, and consequently appears 3.76 times fainter than Sirius. Sirius has 2.03 times the mass and 1.80 times the radius of the sun, but it produces energy at a rate nearly 50 times greater. Regardless of its size and brightness, Sirius, for historical reasons, is astronomically classified as a dwarf star. We now know a great deal about Sirius and can calculate its history and its destiny with considerable confidence.

In spite of its celestial prominence, Sirius managed to conceal an important family secret, until some 160 years ago, when it was discovered to possess a strange faint companion. It is the nature of this tiny companion that has commanded most of the subsequent scientific interest in Sirius and which helped to revolutionize our understanding of how stars evolve and finally die. This small star has also provided a natural laboratory that first demonstrated some of the more striking discoveries of 20th century physics. It is the story of this remarkable companion that dominates the latter half of this book.

This book naturally divides itself into five parts. The first section concerns the ancient lore that surrounds Sirius: the ideas and legends of the ancient Egyptians, Greeks, and Romans, as well as those of other cultures such as the Islamic, and the Chinese. The second section begins the long journey towards a scientific understanding of the stars and Sirius, starting with the concepts of the ancient Greeks. This section also includes the critical period, beginning a little less than five hundred years ago, when Copernicus, Kepler, and Galileo helped to overthrow the old Greek ideas of the cosmos. Later in this period key figures, such as Isaac Newton, Edmond Halley, and others, played a prominent role in changing our ideas about the stars, their distances, and their motions. During this time two key developments occurred: the realization of just how far away the stars really are and the discovery of double stars. Both of these developments made possible the appreciation of such stellar properties as their true luminosities and masses. The second section of the book concludes with the fascinating story of how the mysterious companion of Sirius was first theoretically deduced in 1844, and then finally discovered in 1862. The third section of the book follows 19th and early 20th century astronomers as they began using new instruments and techniques to make some the first physical measurements of the stars: including temperatures and radii, and the determination of the chemical composition of the atmospheres of the stars. This section concludes with the highly successful applications of these developments, and traces the new ideas of quantum physics as they were used to determine how the stars produce their energy and how they evolve and die. The fourth section of this book takes an in-depth look at some of the modern mythologies that surround Sirius. These include: the idea that Sirius appeared as a prominently red star some two thousand years ago; that the Sirius system may conceal a third member; the fascinating way that Sirius figures in the cosmogony of the Dogon tribe of west central Africa; and the story of a modern cult that made Sirius a central aspect of their beliefs, with disastrous consequences. The fifth and final section of the book takes an up-to-date scientific look at what new observations of Sirius from space have revealed. It concludes with the stellar evolution of Sirius and its companion, how they appeared in the past, how they will appear in the future, and how this evolution contrasts with that of our sun.

I was drawn to Sirius more than twenty years ago, when I first observed the star with the ultraviolet spectrometer on board the *Voyager 2* spacecraft, while it was traveling between Saturn and Uranus. I was hooked, and have continued making observations of Sirius using other spacecraft ever since. During most of this period, my interest was primarily astronomical and centered on obtaining better estimates of the mass and radius of the white dwarf companion to Sirius. Five years ago, however, I became involved in a project to observe this white dwarf with the *Hubble Space Telescope* in an effort to improve the determination of its orbit. Part of this study involved a recomputation of the orbit, from all of the old visual observations going back over 160 years. This ultimately required collecting thousands of observations scattered through scores of old journals. It proved necessary to visit libraries across the country to locate some of the more obscure older references. Through this process I became aware of the scope of the history of Sirius and began to appreciate how influential the star and its companion have been in the development of modern stellar astrophysics. It seemed to me that the only adequate way to tell this story was to go all the way back to the beginning, to ancient Egypt, and conclude with an up-to-date account of what we are still learning today about this remarkable star. I hope that this book accomplishes these objectives.

Jay Holberg

NOTES ON TERMINOLOGY, UNITS AND DATING CONVENTIONS

Astronomy, with its huge distances, enormous masses, and unfathomable time spans, faces a problem in conveying its discoveries to the general public. For the distances to the stars, I use variously, the light year (the distance that light travels in one year), and the parsec (the distance at which the diameter of the earth's orbit would appear to subtend an angle of one arc second). (One parsec = 0.306 light years = 3.09×10^{18} centimeters = 1.92×10^{13} miles, where in scientific notation, the superscripts following the ten indicate the number of trailing significant zeros in the quantity.) In discussing the stars I have tried, wherever possible, to use relative units, such as the mass, radius, and luminosity of the sun. Elsewhere, when discussing other aspects of the stars I follow the astronomical practice of using the centimeter–gram–second version of the metric system. In these units the solar mass is 1.99×10^{33} grams, the solar radius is 6.96×10^{10} cm, and the solar luminosity (its total power output) is 3.85×10^{33} ergs/second $= 3.85 \times 10^{26}$ Watts. In addition to the parsec and light year astronomers have other unique units of measurement, which are ingrained in the field. Some of these are the Angstrom unit for measuring the wavelengths of light (1 Å $= 10^{-9}$ m $= 3.94 \times 10^{-10}$ inches); for example, the wavelength of the red line of hydrogen is 6656 Å. In the solar system the convenient unit of distance is the mean distance of the earth from the sun, the Astronomical Unit (1 AU $= 1.50 \times 10^{8}$ km $=$ 93 million miles). Stellar magnitudes are another area of potential confusion. Magnitudes are basically negative logarithmic measures of stellar brightness, so that the brightest stars have the lowest magnitudes and the faintest the highest. The basis of this system is Vega, defined to be magnitude 0. The step size of 1 magnitude is a factor of 2.51 in brightness. Thus, Sirius with a magnitude of −1.44 is 3.77 times *brighter* than Vega, while the faint companion of Sirius, at magnitude 8.48, is 2470 times *fainter* than Vega. For other potentially unfamiliar, or technical, terms there is also a Glossary (see Appendix A, p. 225).

Mathematics is the native language of the physical sciences, but it creates barriers for some readers. For this reason I have endeavored to make limited use of a few simple equations, such as proportions and inverse relations that are needed to express important concepts. Wherever practical, this usage is accompanied by written statements of the relations being expressed. For more complex relations, I have tried to use graphs and figures. I also presume the reader is familiar with the shorthand of exponents, and logarithms for dealing with large numbers.

For familiar historical reasons, I have tended to use English units when discussing instrumentation, prior to the 20th century; for example, in referring to telescope apertures and focal lengths, such as the Clark's $18\frac{1}{2}$-inch 1862 refractor. Dates and spelling are another issue. I use the traditional Latin eras, BC and AD, to designate historical dates, and use Julian calendar dates prior to the introduction of the Gregorian Calendar in 1582, unless otherwise noted. For Egyptian dates and spellings I have adopted the conventions of the *Oxford Encyclopedia of Ancient Egypt*.

Figures

Abbreviations and Acronyms

ANS	Astronomical Netherlands Satellite
AU	Astronomical Unit
CCD	Charged Coupled Device
DA	A white dwarf having a pure hydrogen atmosphere
EUV	Extreme UltraViolet
EUVE	*Extreme UltraViolet Explorer*
FUSE	*Far Ultraviolet Spectroscopic Explorer*
HST	*Hubble Space Telescope*
LHS	Luyten Half arc-Second stars
LTT	Luyten Two Tenths
NICMOS	*Near Infrared Camera and Multi-Object Spectrometer*
ROSAT	*The Roentgen Satellite*
STIS	*Space Telescope Imaging Spectrograph*

Part One

Ancient Sirius

More than any other star, with the obvious exception of the sun, Sirius has fascinated and intrigued humans. This fascination extends far back into the prehistory and antiquity of many cultures. The two main influential tributaries of this ancient stream of myth and legend are those of Dynastic Egypt and the later Greek and Roman civilizations. For the Egyptians, Sirius was quite literally an aspect of the goddess Isis, who played a key role in their mythology and religion. On a more practical level they used Sirius to regulate the Egyptian calendar and to serve as a harbinger of the annual flooding of the Nile River. For the Greeks and Romans, Sirius was never a god, like Mars or Jupiter, but rather was associated with a number of distinct beliefs and ideas. Some of these are still part of our culture and language and even now and then manage to inspire intriguing lines of contemporary scientific inquiry. The first two chapters of this book are a guide to the sources of these ancient beliefs and lore, from their earliest origins in prehistory up to the end of the Roman Empire. What makes this part of the journey especially interesting is how seemingly ancient beliefs, ideas, and discoveries about Sirius and the stars reappear in later eras in the most unexpected ways.

1

The Goddess of the Eastern Horizon

> *"Lady of stars"*
> *"the great one in the sky, ruler of the stars"*
> *"who brightens the Two Lands with her beauty"*
> —Egyptian Temple Inscriptions

DAWN—JULY 17, 2276 BC—MEMPHIS, EGYPT

In the early morning darkness an Egyptian priest is hurrying to the place he comes every morning at this time of year. Memphis, the capital of the first dynasties of ancient Egypt, sits on the west bank of the Nile River, occupying a flat plain covered by palms and tall papyrus reeds, near the point where the river begins spreading into its broad delta. The priest's destination is a location that has been carefully chosen to provide a clear, unobstructed view of the eastern horizon. He and other priests come to this place, in the pre-dawn hours, during this particular season each year, to watch the sky and wait. By the time the priest arrives, Re, the sun, is still well below the horizon. Although the eastern sky is beginning to brighten, the priest can easily see the stars of Osiris hanging above the horizon. His attention, however, is focused on a distant point, just above where the sky meets the low hills of the eastern desert, far beyond the quietly flowing Nile. He has been watching carefully for perhaps fifteen minutes before he imagines he sees it. At first the star Sopdet is a small, almost imperceptible, point of light very low in the sky. Initially it is a faint reddish-gray color but it quickly brightens, turning a pale yellow, as it rises. Sometimes Sopdet appears as a point of light, but frequently it appears to disintegrate into a cascade of color and almost disappears before reassembling itself. The entire event lasts only a short time before the approach of Re brightens the sky making it hard to see Sopdet, even when the priest knows exactly where to look.

The dramatized events of that particular morning, almost 48 centuries ago, can be imagined from what we have learned of Egyptian history, religion, and astronomy. Sopdet was the star Sirius, and the celestial manifestation of the goddess Isis. The stars of Osiris, god of the Nile and rebirth, are part of the constellation Orion that rises just prior to Sirius. Isis and Osiris had once again returned from the *duat*, or underworld. Their journey had begun several months before: when they "died" in the western evening sky and now they were being "reborn" in the eastern morning sky. The advent of Sopdet's annual appearance meant that the god Khnum, lord of the cataract near Aswan, would now open the gates of his cave, releasing a torrent of water, and soon the Nile would begin to rise, flooding, and fertilizing the fields as it did each year.

Once the events of that morning were reported, other priests would open the inner sanctums of the temples and the statues of the gods would be made ready. The great festival would be held, sacrifices would be offered, long solemn religious processions of the clothed gods would be conducted and hymns would be sung. Sopdet returned every year at this time, but on this particular morning, of this particular year, she had arrived on the first day of Thoth, the first month of the Egyptian calendar and the day when the new year was celebrated. The first rising of Sirius and the new year coincided only once every 1461 years, and according to the Egyptian calendar, a great cosmic cycle had been completed and a new one begun. It was the Egyptian equivalent of a new millennium.

The celestial pageant witnessed by the priests on that long-ago morning is known today as the heliacal rising of Sirius, the first day when Sirius reappears in the morning sky after seventy days of absence, during which it is lost in the glow of the sun. The heliacal rising (meaning, to appear with the sun) occurred each year just prior to the annual flooding of the Nile and the two events were very much linked in the minds of the people of that time. The star Sirius was intimately connected with Egyptian religion and mythology, the ancient Egyptian calendar and time-keeping, as well as burial practices, and temple construction. Even today, the way that the ancient Egyptians determined the advent of their new year, with respect to the rising of Sirius, plays a major role in helping historians and archeologists to determine absolute historical dates for the pharaohs and dynasties of ancient Egypt.

The ancient Egyptians had a very special relationship with the star Sirius. Sirius (Sopdet) was the heavenly representation of Isis, who in the Egyptian pantheon was the daughter of Re and the sister and wife of the god Osiris. Osiris in turn was associated principally with the stars of Orion's belt. In the sky, Sirius is situated to the south and east of Orion, and the stars in Orion's belt appear to point toward Sirius. Sirius was often depicted as Isis, standing in a celestial boat with a prominent five-pointed star over her head, and facing Osiris on the right standing in his boat (Figure 1.1).

Osiris was lord of the underworld, whom the departed sought to join in afterlife. He was also the embodiment of the Nile River. The most enduring of all Egyptian myths tells how Osiris was killed by his jealous brother Seth, who dismembered Osiris and scattered his remains throughout the length of Egypt. Horus, the son of Isis and

Figure 1.1. Isis and Osiris in their sacred barges and below their stellar associations. From the Ramesseum, the mortuary temple of Rameses II at Thebes (Michael Soroka).

Osiris, defeated and banished Seth. Isis then wandered through Egypt, recovering and reassembling the body of Osiris, who was subsequently resurrected. This tale of death and rebirth resonated strongly with Egyptians, for whom it was a powerful metaphor for the annual renewal of the crops, and for the yearly cycle of the rising and falling of the Nile, on which the crops depended.

Isis was associated with the giving of life, the fecundity of nature and nurturing. She was also associated with resurrection and funerary rites, in particular the ceremony of the "Opening of the Mouth". This ceremony was performed by the priests on the fully embalmed mummies as a last right of passage, and was intended to reanimate the body so that the deceased might nourish themselves during the afterlife. When the mouth of the deceased was ritually opened they could once more live. The seventy-day period when Sirius was invisible, and during which Isis and Osiris passed through the *duat*, paralleled the seventy days the Egyptians traditionally used to prepare a mummified body for burial. Isis' role in resurrecting Osiris has an obvious heavenly association with the annual reappearance of the star Sirius, preceded by the stars of Orion in the morning sky. The symbolism of the annual cycle of the crops, the rhythm of the Nile, and the associated rebirth and renewal was compelling to Egyptians. It was the reappearance of Sirius each summer that heralded the new year and beginning of

Figure 1.2. Sirius as the goddess Hathor. A depiction of the figure of Sirius/Hathor from the Circular Zodiac of the Temple of Hathor at Dendera (Michael Soroka).

the cycle of the river and the crops. In this respect, Isis was referred to as "the mistress of the beginning of the year" and the "goddess of the eastern horizon of the sky, at whose sight everyone rejoices".

The goddess Hathor was also associated with Sirius. Hathor, whose name means "House of Horus" is often depicted as a cow or cow-headed goddess and was alternatively believed by Egyptians to be the mother of Horus and the daughter of Re. Hathor, like Isis, was a goddess of love, but a love more focused on fertility than on the maternal aspects associated with Isis. Hathor received Re in the west each evening, protecting him during his journey through the underworld until he was reborn in the morning sky. A common image of Hathor is a standing cow-headed goddess with the disk of the sun cradled between her spreading horns. In other images, a nocturnal Hathor is a fully bovine kneeling cow, with the star Sirius between her horns (Figure 1.2).

By Greco-Roman times, a center of Hathor worship was firmly established at Dendera, where a well-preserved temple of that era still exists on the western bank of the Nile, some fifty kilometers north of Thebes. The principal temple at Dendera was dedicated to Hathor, but there was also a smaller adjacent temple dedicated to Isis. Dendera was the reputed birthplace of Isis, and there an important festival was held on the first day of the Egyptian calendar, when the statue of Hathor was brought out onto the roof of the temple to be reunited with Re and the rays of the sun.

The worship of Isis has a long, well-established history throughout ancient Egypt. Later, beginning in the first century AD, Isis cults became prominent over much of the Roman Empire, even reaching as far as London. The 2nd century BC Ptolemaic rulers and the 1st century AD Roman emperors remodeled and made additions to several Isis-related temples, such as those at Dendera and Philae. As late as the sixth century of the Christian era in Egypt, the temple of Isis at Philae near Aswan was still in use. The relation between Isis worship, the night sky, and Sirius is prominently evident in a number of Egyptian temples, particularly those at Dendera and others near Aswan. The splendid temple of Hathor at Dendera had a beautiful vaulted ceiling with a circular *bas relief* representation of the heavens, with depictions of the constellations, the zodiac, and the planets (Figure 1.3). The original ceiling of caramel-colored

Figure 1.3. The Circular Zodiac at Dendera. Sirius as the kneeling cow-goddess Hathor indicated by the arrow (*An Historical View of Hindu Astronomy*, John Bently, 1825).

sandstone was removed in the 19th century and now resides in the Louvre Museum in Paris. On this ceiling, "the Circular Zodiac", the vault of the heavens is supported by a ring of carved figures of a standing Isis and a kneeling Horus. Among the figures representing animals and gods on the ceiling we can recognize the familiar signs of the zodiac such as Gemini, Aries, Cancer, and Sagittarius. These constellations are clearly of later Persian and Greek origin and are not strictly Egyptian. Less familiar, but more clearly Egyptian, is the bovine image of a kneeling Hathor, sailing the sky in her celestial barque with the star Sirius between her horns. Directly behind Sirius/Hathor is an image of a female with a bow and arrow. This is Satet who, in some myths, is responsible for releasing the flood waters of the Nile.

At the end of the 19th century, the British astronomer Sir Norman Lockyer visited Egypt to study the astronomical associations of the Egyptian monuments. Lockyer, regarded by many as the father of archaeoastronomy, was one of the first to suggest that many ancient monuments, including certain Greek temples as well as Stonehenge in England had astronomical associations. In 1890, on a trip to Egypt, he made first-hand studies of numerous monuments. This visit resulted in his influential 1894 book, *The Dawn of Astronomy*. In it Lockyer claimed to have established that a number of monuments were deliberately oriented to be in alignment with astronomical events such as the direction of sunrise at the solstices. In particular, he maintained that the Temple of Isis at Dendera had been constructed in deliberate alignment with the direction of the heliacal rising of Sirius. However, because of the precessional shifting of the earth's pole (see Appendix A), such an alignment would persist for at most a few centuries before the growing misalignment would become noticeable. The present Temple of Hathor was built near the end of the Ptolemaic era, around the time of the Roman conquest in 30 BC. However, the foundations of earlier temples also exist at the site with offset alignments from the Ptolemaic era temple. Lockyer maintained that these earlier temples also were aligned with respect to the shifting location of the heliacal rising of Sirius and in fact constituted a sequence of structures built to maintain this alignment.

Another temple dedicated to Isis exists on the island of Philae in the Nile at Aswan. The nearby temple of Satet also appears to have both calendrical and religious significance associated with Sirius. The Egyptians may well have attempted to calibrate the flooding of the Nile at this temple. Although most of the present temples are from Greco-Roman times, they still convey much of the religious mystery and ceremony associated with Isis worship and its association with Sirius.

To an observer on the earth's surface, the sun has two apparent motions. The first is its daily east to west journey across the sky due to the rotation of the earth about its axis. The second motion is subtler and involves the sun's path through the heavens due to the earth's orbit about the sun. Over the course of a year, the sun, reflecting the orbit of the earth about the sun, appears to move eastward with respect to the fixed stars so that each day the stars near the equator appear to rise approximately four minutes earlier than on the previous day. After a year, the sun returns to the apparent position in the sky where it started and the cycle repeats. Any star near the celestial equator will thus appear to approach the sun and become lost in its glare at evening twilight. After a period of time, the star will reemerge in the morning sky. In the case of Sirius, depending on the latitude of the observer, Sirius disappears in mid-May and reappears approximately seventy days later in early August (in our Gregorian calendar). Throughout the ancient world, the annual occurrence of these events—in particular the first reappearance of certain stars—marked the seasons and in some cases had great religious significance. During the early history of Egypt, scholars believe that observations of the heliacal rising of Sirius were conducted near the city of Memphis, or perhaps at the Temple of Heliopolis near modern Cairo. It has been suggested that later in the New Kingdom (1570 BC–1070 BC), observations were shifted to a site some 6 degrees of latitude farther south at the Temple of Satet near Aswan.

The heliacal rising of Sirius refers to the date when the star has finally emerged from the glare of the sun and is visible for the first time just before sunrise. For Sirius, this generally occurs in the latter half of July, the exact date again depending on the latitude of the observer, local observing conditions, and the historical epoch. As a rule of thumb, once Sirius has cleared the thickest layers of the atmosphere by rising to an elevation of approximately 6° above the horizon, and while the sun is still 5° below the horizon, Sirius will be just visible. Dr. Bradley Schaefer of the Louisiana State University has carefully investigated the optimal date for sighting the heliacal rising of Sirius, as a function of time and geographical location, extending back through time to 3500 BC. He even made his own observations of the heliacal rising of Sirius in Egypt at the latitude of the ancient city of Heliopolis. Taking into account the precessional motion of the earth's pole, the relative motion of Sirius and the sun, and estimating the transmission of light by the earth's atmosphere near the horizon, he finds that up until the first century AD the heliacal rising occurred on the date of July 17 plus or minus about one day, in the old Julian calendar. However, given the range of natural variations in the transparency of the earth's atmosphere as well as other factors, the actual date of first visibility during any given year can vary by as much as two days either side of July 17. At present, the date of this event has slipped to about August 4th in our present Gregorian calendar. The Egyptians referred to the heliacal rising of Sirius and its associated festival as *prt spdt*, "the going forth of Sopdet".

We know little of how the Egyptians described the physical appearance of the stars or even of how they observed the stars. There is, however, an account from the Hellenistic Greek astrologer, Hephaestion of Thebes, in approximately 425 AD. He noted that the Egyptians regarded Sirius as having great astrological importance and that they were attentive to the color of the star as it appeared at heliacal risings. If Sirius was white, the Nile flood would be high and the current strong. If it was "fiery red and the color of red ochre there would be war."

Prior to the Old Kingdom, the earliest Egyptian calendars were based on lunar cycles. Because there are, on average, 12.4 lunar cycles in a solar year, some mechanism is required if lunar calendars—such as the Chinese, Jewish, or Islamic calendars—are to be kept in phase with the seasons. Often this involves periodically adding an extra month when required. In Egypt, the heliacal rising of Sirius served the purpose of synchronizing their lunar calendar with the solar year. If the heliacal rising of Sirius occurred in the last eleven days of the lunar calendar the priests added an extra month. The Egyptians recognized three seasons centered on the flow of the Nile and the agricultural cycle: *Akhet* or Inundation, *Peret* or Planting and Growth, and *Shemu* or the Harvest and low-water season. Each season was exactly four Egyptian months long. During the Inundation season which coincides with our months of June to September, the Nile River swells with the monsoon rainfall that originates in the Central African and Ethiopian highlands. Before the construction of the modern Aswan High Dam, the crest of rising water would travel the full length of Egypt in about ten days, reaching the lower Nile near the end of July, and finally peaking in mid-September. The beginning of the annual floods thus coincided with the heliacal rising of Sirius and the Egyptians associated the onset of the Inundation with this event.

Sometime during the first or second dynasties, the Egyptians developed a very simple but highly effective calendar based on the sun. This calendar consisted of 12 equal months of thirty days each, with each month further subdivided into three ten-day "weeks". At the end of the year, there were five remaining days, which the later Greeks called "epagomenal" or "added-on days". This civil calendar, which was used for religious and administrative purposes, had a total of 365 days. There were no leap years, leap centuries, or months of differing lengths. Every year was identical to the next, beginning on the first day of the first month and ending exactly 365 days later. Initially the months were simply numbered but later they received names: Thoth, the god of the moon and the reckoner of time, was the first month while Mesore, the last month, was associated with the rebirth of the sun. Until the Roman era, the Egyptian civil calendar followed this precise pattern. This high degree of regularity was virtually unique in the ancient world and readily lent itself to the reckoning of long spans of time. In contrast to other national and regional calendars in use at that time the Egyptian civil calendar was relatively unambiguous, and continued to be used well past pharaonic times for the calculation of astronomical dates. For example, as late as 1530 AD, the Polish astronomer Copernicus used the Egyptian civil calendar as a means of converting various astronomical dates into a common framework. The chief disadvantage of the Egyptian civil calendar was that it contained no provision for leap years so that the tropical year—the time from one summer solstice to the next—was slightly longer than the civil year. The tropical year is almost $365\frac{1}{4}$ days long and therefore the Egyptian civil calendar lags the tropical year at a rate of one day in four years and consequently the calendar slips with respect to the seasons. It has been suggested, by the American archaeoastronomer Dr. Ronald Wells, that the upper Nile temple of Satet was built to astronomically determine the 5-day difference between the civil and solar years. He notes that during the time of the construction of the temple, in the reign of Queen Hatshepsut around 1450 BC, the heliacal rising of Sirius would have occurred five days after the summer solstice, providing an observational link between the two calendars. The location of this temple, at a latitude of 24°, and its orientation were intended to take advantage of this unique coincidence during that era.

Our present calendar has its roots in the Egyptian civil calendar. When Julius Caesar reformed the Roman calendar in 45 BC, he employed the Greco-Egyptian astronomer, Sosigenes, to devise a solar calendar of $365\frac{1}{4}$ days in length through the addition of one extra day every fourth year, the leap year. This *Julian* calendar served the Roman and Christian worlds for nearly 16 centuries. However, by 1582, it was out of phase with the seasons by 11 days. One result was that the date of Easter began increasingly to lag the spring equinox. This occurred because the actual solar year of 365.2422 days is 11 minutes 23 seconds shorter than the Julian year of 365.25 days. Thus, it loses about three days with respect to the seasons every 400 years. Pope Gregory VIII solved this problem by decreeing, that following Thursday, October 4, 1582, ten days would 'disappear' and the calendar would jump to Friday, October 15, 1582. This calendar reform also adopted the convention of dropping the leap year on those century years not evenly divided by the number 400, for example the years 1800 and 1900 were not leap years but the year 2000 was. This device keeps our *Gregorian* calendar in phase with the seasons over several millennia. The adoption of the Gregorian calendar encountered religious and political objections in the non-Catholic world and was not made official in England (and the North American Colonies) until 1752 and not in Russia until 1918.

Because the Egyptian civil calendar slips approximately one day every four years with respect to the seasons, the heliacal rising of Sirius also becomes further and further out of phase with the civil calendar in successive years. When this slippage amounts to a full year, the heliacal rising of Sirius will return to the same day with respect to the civil calendar. This occurs after approximately 1461 years ($4 \times 365.25 = 1461$). Thus, a period of 1461 Egyptian years is equal to 1460 Julian years. This period of time is known as the "Sothic Cycle" for Sothis, the Greek name for Sirius. We know from the 3rd century AD Roman author Censorinus that the heliacal rising of Sirius coincided with the 1st day of Thoth (July 20), in 139 AD. Working backwards from this date, through previous Sothic Cycles, we find that the heliacal rising of Sirius coincided with Thoth 1 during the New Kingdom (1570 BC–1070 BC) in about 1332 BC and in about 2782 BC, prior to the beginning of the Old Kingdom (2686 BC–2180 BC). However, as will be evident, this scheme is not as unambiguous as it sounds.

The pyramids at Giza are the most striking remnants of ancient Egyptian civilization. From Egyptian records, we know that the Great Pyramid was built during an 85-year period around the reign of the Pharaoh Khufu in Dynasty Four of the Old Kingdom. The exact time range to which this corresponds in our modern Gregorian calendar is difficult to establish with confidence. The best estimate is: sometime between 2589–2504 BC. People in the ancient world almost invariably referred to the year in which an event took place by stating that it occurred so many years into the reign of a ruler or local magistrate. For example, the Roman philosopher Seneca writes of a devastating earthquake that damaged the city of Pompeii in 62 AD, as having occurred during the year of the consulship of Regulus and Verminous. To relate such dating schemes to a modern calendar date, an accurate list of rulers and their reigns is required. To complicate matters further, different nations and even cities reckoned dates according to local conventions. One very useful list of rulers was produced in the 2nd century AD by the Alexandrian astronomer Claudius Ptolemy. This list begins with the Babylonian kings of the 8th century BC and continues through the Persian kings and a series of Hellenic rulers, and finally the Roman Emperors, up to Hadrian. Armed with this list and a tabulated cumulative year count, it is possible to determine the number of years between any two events during this period.

Historically, scholars have developed dating schemes for ancient Egypt by working backwards from Roman and Greek times, through the Egyptian dynastic lists and the lengths of the reigns of the many pharaohs, to arrive at relative calendar dates for a chronology of ancient Egypt. One of the pitfalls of this approach is the reliance on the fidelity of ancient lists, particularly for periods when effective government had broken down. Simply put: there is no assurance that the royal genealogies are correct. One method that scholars use to constrain and verify ancient Egyptian chronologies is to use the Sothic Cycle to establish specifically dated benchmarks during earlier pharaonic times. Beginning with an historically fixed date—such as the Censorinus statement that the heliacal rising of Sirius and 1 Thoth, the Egyptian New Year, coincided in 139 AD—it is in principle possible to establish an absolute calendar date for any similar reference to a heliacal rising of Sirius which can be clearly associated with a civil calendar date in the reign of a particular pharaoh. For example, there exists a letter telling a priest that the heliacal rise of Sirius will occur on day 16 of month 8 of the year 7. The year refers to year 7 of the reign of pharaoh Sesostris III of the 12th Dynasty of the Middle Kingdom (2040–1782 BC), which corresponds to a date of 1866 BC. For several other documents it is possible to make similar associations, all in the New Kingdom. However, there are important caveats concerning the use of such benchmarks to establish accurate absolute ages. For example, it is uncertain exactly how or where the observations of the heliacal rising were made. Were they made in the old capital of Memphis, or farther south in the later capital of Thebes, or even near Aswan? For every degree south from Memphis, Sirius rises one day earlier, and each day of change in the date of the heliacal rising produces a corresponding shift of 4 years in the length of the Sothic Cycle. Taking into account reasonable estimates for all uncertainties, including adopting a Sothic Cycle of 1457 ± 20 years, Dr. Schaefer estimates that the Sothic Cycle prior to 139 AD started in 1319 BC plus or minus 20 years. Although very useful in helping to fix dates as far back as the Middle Kingdom, Sothic dating generates considerable scholarly debate and contending chronologies.

Our modern practice of dividing the day into twenty-four hours has it origins in the methods the ancient Egyptians used to determine time during the night and their practice of dividing the night into twelve hours. In order to measure time at night, when sundials were of no use, the Egyptians used "star clocks". These consisted of lists of 36 bright stars or small groups of stars called *decans* that were distributed more or less uniformly around a wide, generally southern, equatorial zone of the sky. Because of the earth's orbital motion, these *decans* would be observed to rise sequentially during a given hour of the night at roughly ten-day intervals during the course of the year. Each *decan* was associated with one of the 36 ten-day "weeks" of the Egyptian calendar. It is not clear exactly which stars or star groups the Egyptians used for this purpose, but Sirius and the three stars of Orion's belt were certainly among them.

Examples of star clocks have been found on the interiors of coffin lids and tomb ceilings dating from the Middle Kingdom. Basically, they consist of tables with columns corresponding to the 36 decans (and 4 extra places) and rows for each of the 12 Egyptian months. The star clocks required selecting a month and noting the sequence of 12 decans in the sky that were visible each night during that period. Star clocks were far from precise. Since the decans were unequally distributed and spaced in the sky, the "hours" could vary in length during the night and also during the seasons, when the length of the night changes. Additionally, because of the slow precession of the earth's axis, the star clocks would slip out of phase with respect to the sky over a period of 200 to 300 years and have to be redefined.

There exists an interesting connection between the Sothic Cycle, Sirius, and the legend of the Phoenix, a mythical bird that rose from its own ashes. There are many versions of the Phoenix myth but most have elements of the following similar theme. A great bird with fabulous red and gold plumage lives far to the east. After a long time, centuries or millennia, it flies to the temple of Heliopolis in Egypt, where it builds a funeral pyre and is consumed by flames. The Phoenix then rises from the ashes and returns to the east to repeat the cycle. One particular version of this story was discussed by the American astronomer T. J. J. See in 1892. At the time, See was completing his graduate studies in Berlin and had become fascinated by the ancient accounts of Sirius having once been a red star (a detailed discussion of his advocacy of a Red Sirius is taken up in Chapter 11). See pointed out that the Roman historian Tacitus (AD 55–120) had stated in his *Annuals* that the Phoenix returned to Heliopolis every 1461 years and that its last appearance in Egypt was around 34 AD. Tacitus acknowledges that there is some dispute as to the lifespan of the Phoenix, the tradition being 500 years. Leaving aside Tacitus' report that the Phoenix, actually "appeared" in Egypt in 34 AD, the use of the precise figure of 1461 years leaves little doubt that there existed some Egyptian tradition of identifying the return of Phoenix with the Sothic Cycle and the heliacal rising of Sirius. See also used various associations, such as the bird returning from the east, and in particular the red color of its plumage, to identify the origin of the Phoenix legend with Sirius.

Retired Rutgers physics professor Dr. Michael Molnar has called attention to a more concrete and fascinating link between the Phoenix and the Sothic Cycle. Among the many commemorative coins issued in Alexandria during the reign of the Roman Emperor Antoninus Pius, were large tetradrachmas depicting the emperor on one side and on the reverse a Phoenix crowned with a halo above which was the Greek word "aeon", which was meant to herald the arrival of a new "great age" or era (Figure 1.4,

Figure 1.4. A tetradrachma from Alexandria from the reign of the Roman Emperor Antoninus Pius (left), showing the Phoenix crowned with a halo (Milne 1737; Courtesy of the Michael R. Molnar Collection).

see also color section). The date of the issue of the coin was the second year of the emperor's reign or 139 AD, which coincides with the beginning of the new Sothic Cycle. A similar coin was issued four years later. Molnar interprets the second coin as a consequence that the heliacal rising of Sirius would have occurred on Thoth 1 over a period of four years, since the Sothic calendar and the Julian calendar only drift apart by one day in four years.

Phoenix is the Greek name for the bird which the Egyptians referred to as the *bennu*, that they associated with the soul of Re. The *bennu* guided the gods on their journey through the *duat* until they were reborn in the morning sky. The *bennu* was said to alight on top of the *benben*, a sacred conical or pyramidal-shaped stone that resided in the temple of Heliopolis. In Egyptian creation mythology the *benben* was the first hillock of dry land to emerge from the primeval waters. This association is evident in the hieroglyphic symbol for Sirius:

The symbol above the star is that of the letter "T", indicating the feminine form of the word star, while the triangular symbol is a pyramid, or perhaps an obelisk or the *benben*. In the celestial context, the *bennu* died annually in the glow of the evening sun and was reborn in the eastern morning sky some seventy days later.

The Egyptians maintained one of the oldest, most enduring, and coherent of all civilizations. Their outlook was focused on eternity and the natural rhythms of life along the banks of the Nile. Although they were careful observers of the heavens their curiosity about what they observed was limited for the most part to the mythological pageant that passed overhead from night to night and year to year. The study of the heavens served the needs of the Egyptian religion and provided a means of reliable time-keeping. Advancing beyond these achievements, however, would require the infusion of new ideas.

These new ideas began arriving in the 6th century BC as the cultural baggage of a sequence of foreign armies that successively conquered Egypt. The first to arrive were those of the Persian Empire under Cyrus the Great, who conquered Egypt in 674 BC and made it a Persian satrapy. This was followed in turn by the army of Alexander the Great in 332 BC, who defeated the Persians and established the Ptolemaic empire. The final conquest of ancient Egypt was that of the Roman Empire. It began with Julius Caesar in pursuit of his strategic alliance with Queen Cleopatra, and ended with his nephew Augustus Caesar who achieved the title of Roman Emperor by defeating his rival Mark Antony and his consort Cleopatra. Egypt finally became a Roman province in 30 BC. These conquests effectively ended Egyptian independence after 30 Dynasties of Pharaohs, but they also spread Egyptian culture across the Mediterranean world and helped launch a scientific revolution that peaked in the 1st century AD and then almost died with the end of the Roman Empire.

2

The Dog Star

"The ancient Greeks have a knack of wrapping truths in myths"
—George Lloyd (1913–1998)

Among the brightest stars of the northern winter sky, Sirius is prominent as the principal star of the constellation Canis Major, Latin for the Greater Dog. Canis Major is located below and to the left of Orion, the Hunter. Above Canis Major is Canis Minor, the Lesser Dog (Figure 2.1). Both of these Dog-Constellations represent Orion's faithful hunting dogs and constant companions, who track and fetch his prey. Indeed, Canis Major, at the foot of Orion, is often envisioned as being poised, ready to spring towards the constellation of Lepus, the Hare. Canis Minor also contains its own prominent dog-star, Procyon, whose name in Greek literally means the one that goes before, or precedes, "the dog". This refers to the fact that Procyon is seen to rise approximately an hour before Sirius. Together, Sirius, Procyon, and Betelgeuse, the, brightest star in Orion, form the three stars of the "Winter Triangle". The well-known relations between Sirius, Canis Major, Canis Minor, and Orion are a familiar part of our sky-lore, much of which comes from early Greek sources. Indeed, Canis Major, Canis Minor, and Orion are among the 48 constellations recognized by the early Greeks. However, for Sirius the star, there existed a host of complex associations and beliefs, among the early Greeks and Romans and other cultures, which are both extensive and ancient.

The precise origin of the name Sirius itself is obscure, it is certainly not Arabic, like many of the most familiar stars in the sky, such as Vega and Deneb, nor is it strictly Greek, such as the stars Arcturus or Procyon. The early Greeks often simply referred to Sirius as the "the dog" or "the dog star". The name Sirius (Σείριος) first appears in the 7th century BC, in the poem *Works and Days* by Hesiod. Although the name Sirius may well be associated with the Greek word σείριος ("*sierios*"), meaning searing or burning, the ancient Greeks themselves were uncertain as to the source of

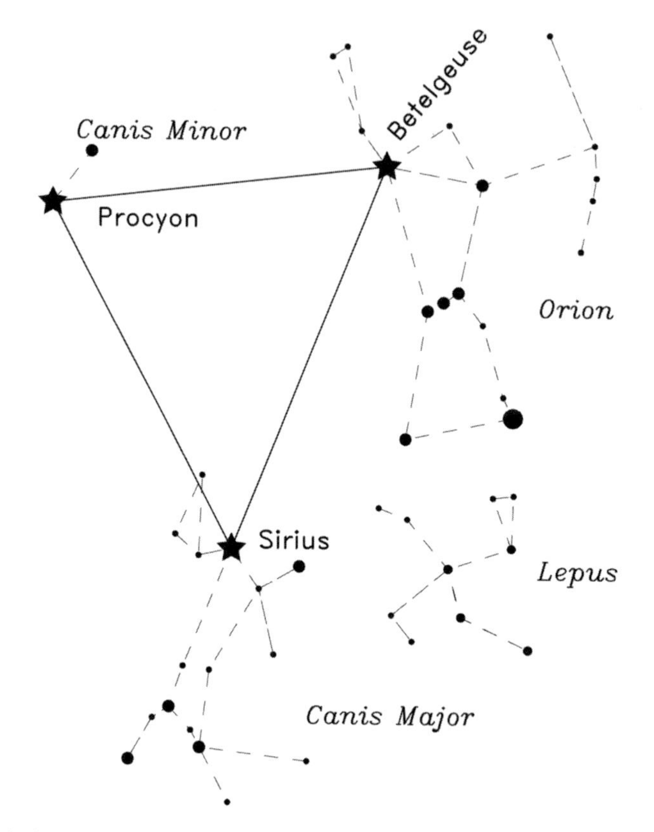

Figure 2.1. A view of the winter sky showing the relationship of the constellations Canis Major, Canis Minor, Lepus, and Orion together with the locations of Sirius and Procyon. The Winter Triangle (solid lines) consists of Sirius, Procyon, and Betelgeuse.

the name. It is very likely pre-Greek, possibly of Indo-European origin. This is suggested by the sacred Hindu texts, the Vedas, which originated in about 1500 BC, and refer to Sirius as *Tishiya* and by other variations that are also echoed in the ancient Iranian *Tišhtrya*. If these names bear any linguistic relation to the word Sirius, then the origin of the name is indeed very old and its genesis is lost in the prehistoric movement of peoples of the Near East and eastern Mediterranean.

The Greeks possessed an elaborate lore associated with Sirius, quite distinct from that of the Egyptians. To the Greeks, the first appearance of Sirius in the morning skies during the final days of July and early August indicated the arrival of the sweltering heat of late summer. Specifically, in the Greek mind the star was also associated with heat, fire, and even fevers. There was also a strong association with dogs, and in some instances with the ominous presence of doom. The earliest recorded mention of Sirius is from the 8th century BC in Homer's epic poem of the Trojan War, the *Iliad*. Although Homer does not explicitly use the name Sirius, he employs all of the common Greek attributes of the star in three poetic metaphors describing the combat of both Greek and Trojan heroes. As translated by Robert Fitzgerald, Homer describes the shining armor of the Greek warrior Diomedes, as the goddess Athena prepares him for battle, and likens it to the rising of Sirius, in book 5 of the *Iliad*:

> Now Diomêdês' hour for great action came.
> Athêna made him bold, and gave him ease
> to tower amid Argives, to win glory,
> and on his shield and helm she kindled fire
> most like midsummer's purest flaming star
> in heaven rising, bathed by the Ocean stream.
> So fiery she made his head and shoulders
> as she impelled him to the center where
> the greatest number fought.

In book 11, Homer uses a darker, more ominous, Sirius to describe the armor of the Trojan hero Hector and to foretell his fate:

> Hektor moved forward with his round-faced shield.
> As from night clouds a baleful summer star
> will blaze into the clear, then fade in cloud.
> So Hektor shone in front or became hidden
> when he harangued the rear ranks—his whole form
> in bronze aflash like lightning of Father Zeus.

Perhaps the most famous description occurs later in book 22, when Homer again uses Sirius to describe the glittering armor of the Greek champion, Achilles, as he prepares for his decisive battle with Hector before the walls of Troy:

> And aging Priam was the first to see him
> sparkling on the plain, bright as that star
> in autumn rising, whose unclouded rays
> shine out amid a throng of stars at dusk—
> the one they call Oríôn's dog, most brilliant,
> yes, but baleful as a sign: it brings
> great fever to frail men. So pure and bright
> the bronze gear blazed upon him as he ran.

In all of these passages Homer deftly uses the brilliance of Sirius as a metaphor to describe the appearance of polished Bronze Age armor. However, these lines also allude to several themes that continually reappear over the next thousand years in Greek and then Roman literature, which reveal popular ideas and beliefs concerning Sirius. The first is the association of Sirius' "flaming" brightness with fire and its annual appearance in the dawn sky, marking the arrival of late summer. The second is clearly different, Sirius is a "death star" used by Homer to convey the fate of doomed Hector in his battle with Achilles. The third reference again describes the appearance of Bronze Age armor, this time that of the victorious Achilles. But to this is added the association of Sirius with dogs, and fever and again as a sign betiding woe. The scene Homer paints of the stars of Orion rising in the night skies of autumn, followed by Sirius, would have been intimately familiar to the Greeks as a seasonal marker. The allusion to dogs is straightforward: Sirius is the chief star of the constellation Canis Major. Indeed, later Greeks, such as the astronomer Ptolemy, often used the simple term "the dog" (κύων) to refer to both the star Sirius and the constellation of Canis Major. The mention of fever, as we shall see, is another common Greek attribute of Sirius.

The *Iliad* was originally an epic poem arising from a long oral tradition. It was recited and acted out by untold generations of bards, poets, and storytellers, that long preceded Homer. Homer, who may never have existed as an individual, is traditionally credited with composing the existing versions of the *Iliad* and the *Odyssey*, shortly before 700 BC. The events described in the *Iliad*, to the extent that they have any historical basis at all, depict a decade long siege of the city of Troy by a Greek army, sometime around 1250 BC. Although, the historical facts surrounding the *Iliad* are murky, many of the principle cities, Troy, Mycenae, Tiryins, etc. have been discovered and the armaments and modes of battle described in the *Iliad* appear to be accurate, but none of the major characters have any sound historical counterparts.

The *Iliad* contains many astronomical allusions in addition to those concerning Sirius. The ancient bards, as well as their audiences, would have been as fully familiar with the appearance of the night sky as they were with the byways of their native village. To anyone who has experienced the extreme good fortune of viewing a moonless night sky, with dark-adapted eyes, in a location completely free from artificial lights, there is little doubt that the image of an inky blackness filled with literally thousands of stars constitutes an impressively dramatic spectacle that leaves a lasting impression. There is little perception of depth or distance, and the stars seem almost near enough to touch. To people in Homer's era, when cooking fires and oil lamps were the only artificial lighting, such a spectacle was a common nightly experience. Moreover, navigators, travelers, farmers, and herdsmen of that time were of necessity, first-hand experts on the stars, the constellations, and their movements. Such scenes, and much more detail, would have been intimately familiar to the Greeks of Homer's time.

Some seven centuries after, the Roman poet Virgil amplifies Homer's use of Sirius in his *Aeneid*, the Roman national epic. Virgil uses the imagery of a comet and Sirius to describe the helmet and shield of Aeneas: "... even as when in the clear night comets glow blood-red in the baneful wise; or even as fiery Sirius, that bearer of drought and pestilence to feeble mortals, rises and saddens the sky with baleful light."

Beginning with Homer and continuing on for the next thousand years, the Greeks, followed by the Romans, left an elaborate and complex set of ideas and beliefs regarding Sirius. Hesiod, a poet from Boeotia, an area of central Greece northwest of Athens, who also wrote in the time of Homer, spoke of the heat of late summer brought on by the arrival of Sirius. In his poem *Works and Days*, as translated by Hugh Evelyn-White, Hesiod provides a distinctively rural, agricultural view of the seasons and the sky in the 8th century BC:

> But when the artichoke flowers, and the chirping grass-hopper sits in a tree and pours down his shrill song continually from under his wings in the season of wearisome heat, then goats are plumpest and wine sweetest; women are most wanton, but men are feeblest, because Sirius parches head and knees and the skin is dry through heat. But at that time let me have a shady rock and wine of Biblis, a clot of curds and milk of drained goats with the flesh of an heifer fed in the woods, that has never calved, and of firstling kids; then also let me drink bright wine, sitting in the shade, when my heart is satisfied with food, and so, turning my head to face the fresh Zephyr, from the everflowing spring which pours down unfouled thrice pour an offering of water, but make a fourth libation of wine.

Four hundred years later, the poet Aratus (c. 310 BC–260 BC) also provided a vivid account of the seasons and their relation to the constellations in his popular poem *Phaenomena*. In a much-quoted passage Aratus, as translated by Douglas Kidd, describes the rising of the constellation Orion followed by the dreaded appearance of Sirius:

> Such is also his guardian Dog, seen standing on its two legs below the soaring back of Orion, variegated, not bright overall, but dark in the region of the belly as it moves round; but the tip of its jaw is inset with a formidable star, that blazes most intensely: and so men call it the Scorcher. When Sirius rises with the sun trees can no longer outwit it by feebly putting forth leaves. For with its keen shafts it easily pierces their ranks, and strengthens some but destroys all the growth of others. We also hear of it at its setting. The other stars lying around about Sirius define the legs more faintly.
>
> Under the two feet of Orion the Hare is hunted constantly all the time: Sirius moves forever behind it as if in pursuit, rises after it and watches it as it sets.

Aratus here refers to the fact that as the constellation Canis Major rises in the eastern sky, below and to the left of Orion, at first only Sirius, among the stars of Canis Major, is visible near the horizon. The word variegated refers to the other stars in the constellation, not specifically to Sirius. As the Greeks delineated the constellation of Canis Major, Sirius variously defines the chin, the mouth, or the snout of the dog. The "Hare" is the constellation Lepus, the hare, located west of Canis Major and south of Orion. Both Hesiod and Aratus mention the heat of late summer that the Greeks believed was actually brought on by the appearance of Sirius. To Hesiod this is simply an enervating heat that is conducive to inactivity. Aratus, on the other hand, depicts Sirius as bringing a dangerous scorching heat that can sere and burn and in particular wilt certain crops. Over the centuries Greco-Roman writers returned to these themes again and again to describe the arrival of Sirius. Aratus' *Phaenomena* was extremely popular and Greek and later Roman writers frequently reworked and commented on his poem. Many of the oft-quoted references to the color, flaming appearance, and vaguely "reddish" properties of Sirius have their origin in this tradition.

To the Greek mind there was a direct causal connection between the arrival of Sirius and the onset of the hot dry days of late summer. Sirius, as it emerged from its conjunction with the sun, was thought to induce the heat and dryness of August. This heat could not only wither plants but influence the behavior of animals as well. Goats would gaze towards Sirius in the east and emit a cry, the wild Egyptian oryx was said to turn towards Sirius and sneeze. People could contract deadly fevers at this time of year, brought on by Sirius; men could weaken during this time and women could be overcome by carnal desire. People, who suffered from the heat of Sirius were said to be "star struck" (*astrobóletus*). Even Hippocrates, the father of medicine, warned of the effects of Sirius.

Sirius was thought to produce "emanations" which could place people and animals in danger of these effects. The idea that Sirius was a source of these emanations could well be linked to the visual appearance of the star when the atmosphere is turbulent and unsettled. At these times the star appears alive and active; seemingly splashing colored rays of light into the sky. Because of its brightness and bluish-white color, Sirius displays such activity much more prominently than other stars and was therefore perceived to be capable of producing effects in humans, animals, plants, and the environment. There was also a widespread association in the Greek mind of the twinkling and flashing of Sirius with such physiological conditions and states as seething, shaking, emptying, and oppression: as if the star was in distress and spewing its light about the sky. Indeed, Sirius acquired such epitaphs as "the Shaker".

During the Hellenistic era the inhabitants of the small Greek Island of Ceos celebrated an important local festival. In late summer sacrifices were offered to the star Sirius and to Zeus to bring the cooling breezes that relieved the heat. There on hilltops of Ceos, the islanders, clad in their armor, would observe the heliacal rising of Sirius seeking signs or omens foretelling the possibility of epidemics during the coming year. If the star rose clear and brilliant, then the prospects for the health of Ceos' inhabitants were good. If the star appeared faint or misty, then its exhalations could prove pestilential. So important was Sirius to the citizens of Ceos that they imprinted their coins with an image of a star or a dog emanating spiked rays (Figure 2.2, see also color section).

The association of Sirius with dogs is far more elaborate than the simple fact the star resides in the constellation Canis Major. Dogs, of all animals, were thought most affected by the annual reappearance of Sirius. Dogs were believed to suffer at this time of year and their panting was an indication of internal desiccation and excessive dryness. When this occurred, dogs were in danger of becoming rabid and their saliva poisonous. Humans could then become rabid and die from a dog bite. The rapid panting of overheated dogs, with their outstretched tongues, was viewed by the Greeks as a sort of "gaping" behavior and was also associated with the "seething" and "shaking" nature of Sirius. In this fashion, Sirius was sometimes also referred to as "the Gaper".

It is this old association of Sirius with the heat of late summer and with dogs that is the origin of the seemingly enigmatic phrase "dog days" or "dog days of

Figure 2.2. A 3rd century BC coin from the Greek island of Ceos. The reverse (right) shows a dog surrounded by radiant rays (Sear 3079; courtesy of the Michael Molnar R. Collection).

summer". The phrase goes back to Roman times when this season was known as "dies caniculares", the "days of the dog-star", Canicula being the Latin name for Sirius. The "dog days" were generally considered to extend from early July to mid-August.

It was not only the heliacal rising of Sirius that provoked the interest of the Romans, the heliacal setting of Sirius, almost three months earlier, was also noted. Beginning in 238 BC and around the 25th of April each year, the festival of *Robigalia* was held in which white-robed priests would sacrifice a red dog. The point of the sacrifice was to entreat the goddess *Robigo* to protect the wheat crop from developing a rust-colored fungus that spoiled the grain. It was believed that the heliacal setting of Sirius around that time of the year was responsible for emanations that produced the wheat rust.

In other ancient references, Sirius was not seen as a domesticated dog but a canine having a ravenous wolf-like nature. Even as late as the 8th century AD Sirius was depicted in this way with bared fangs, a rampant posture, and shaggy mane (Figure 2.3, see also color section). The association of Sirius with wolves has some interesting and curious parallels, which extend well beyond Greece and Rome. The Greek poet Oppian in the late 2nd century AD wrote of a golden wolf that lived far to the east and haunted the hills of the Taurus Mountains, in what is now south–central Turkey. The golden wolf was a beautiful animal of exceptional prowess, whose teeth could pierce bronze. When Sirius rose, however, the golden wolf retreated to his underground lair until "the heat of the Sun and the baneful Dog Star cease." Sirius could easily intimidate and dominate the fierce golden wolf and force his retreat. Interestingly the Roman poet Manilius also writes of a people who, like those on the island of Ceos, observed the rising of Sirius in the east from the top of Mt. Taurus and sought omens as to the health of the crops and the populace and the prospects of peace and war.

Finally, although the Egyptians had no original tradition of associating Sirius with dogs, the influence of later Greek and Roman beliefs did have an impact. Just as Egyptian gods and funeral practices influenced the Greek and Roman cultures there also developed a set of cross-cultural beliefs associated with Sirius. During the reign of the Roman Antoninus Pius, from 138 to 161 AD, there was an interesting drachma

Figure 2.3. Sirius in the form of a rampant wolf-like dog, from the 9th century Codex Vossianus Latinus manuscript in Leiden (Reprinted with permission from *Sky & Telescope*, June 1992).

minted in Alexandria which showed the Egyptian goddess Isis riding on the back of a dog (Figure 2.4, see also color section).

In addition to Greek and Roman beliefs about dogs and Sirius, there exist many curious similarities with other cultures elsewhere. In China, Sirius is known as *Tsien Lang* or the Heavenly Wolf. The association with a wolf is intriguing considering the wide-spread association of the star with dogs and wolves in the West. It is not known if this is merely a coincidence or an indication of a very old tradition that spans the Eurasian continent. On the other hand, there do exist written cuneiform references to

Figure 2.4. An Alexandrian drachma from the reign of the Roman Emperor Antoninus Pius, showing a representation of the Egyptian goddess Isis astride the Greco-Roman canine representation of Sirius (Milne 2358; courtesy of the Michael R. Molnar Collection).

Sirius from ancient Mesopotamia which also allude to dogs. For example, the word *'Kak-shisha* has been translated variously as "the dog that leads" and "a Star of the South". Later Mesopotamian names include *Kal-bu*, "the dog" and *Kakab-lik-u*, "the star of the dog". The Phoenicians were said to have called it *Hannabeah*, "the Barker".

There are also numerous and intriguing associations of Sirius with dogs and wolves from throughout North America. To the Alaskan Inuit of the Bering Straits, Sirius is the "Moon Dog". When the moon comes near Sirius, high winds will follow. Among the Tohono O'odham of the southwestern deserts, Sirius is the dog that follows mountain sheep, a description that was shared with the Seri who lived to the south along the Gulf of California, in Mexico. To the Blackfoot of the northwestern Great Plains the star was "dog-face". Among the Cherokee, whose ancestral home was the central Appalachian Mountain region, Sirius and Antares are the dog stars that guard the ends of the "path of souls", the Milky Way. Sirius, in the winter sky, guards the eastern end, while Antares, in the summer sky, guards the western end. A departing soul must carry enough food to placate both dogs and pass beyond, or spend eternity wandering the "path of souls". Alternatively, the Pawnee of Nebraska have an elaborate and well-developed mythology tied to the heavens. The Skidi (or Wolf) band of the Pawnee called Sirius the "Wolf Star" and the "White Star". According to Skidi cosmology, Sirius brought death into the world and would escort deceased tribal members along the "spirit pathway" (the Milky Way) to the place of the dead in the south. During times of a sacrificial ceremony, a tribal representative of the White Star would sit in the southwest corner of the lodge to watch over the ill-fated sacrificial maiden. Among other Pawnee, Sirius was the Coyote Star, the trickster. The Northern Osage, of the south–central United States, regarded Sirius as the "Wolf that hangs by the side of Heaven".

This is not to say that Sirius was universally associated with dogs in ancient North America. On the contrary, there are many more tribal legends and associations that have nothing to do with dogs and wolves, but none of these have any identifiable common themes. Nevertheless, the numerous and widespread associations with dogs and wolves that do exist are curious. It is not inconceivable that some of this lore may well have crossed the Bering Straits with one or more groups of ancient people from northeastern Asia during the last Ice Age.

In North America, Sirius also served as a direction marker. In addition to being the Wolf Star of the Skidi Pawnee, Sirius was also revered as "The White Star", which was one of the four god stars that held up the sky. Traditional earth lodges were laid out according to these four stars, which occupied the semi-cardinal directions, with Sirius assigned the southwest lodge pole of the structure.

There have been other proposed alignments with Sirius in the Americas; however, none of these has stood up well to scrutiny. Perhaps the best known example is that of the "Medicine Wheel" in the Big Horn Mountains of Wyoming. In 1974 John Eddy, a scientist at the National Center of Atmospheric Research in Boulder, Colorado published an article in *Science* calling attention to supposed alignments of rock features within the wheel with celestial events such as the heliacal risings of bright stars such as Sirius, Aldebaran, and Rigel. Much of Eddy's evidence was undercut, however, by the fact that he did not include the effects of atmospheric refraction (Appendix A) which serves to displace the points on the horizon where Sirius and other stars will appear. When refraction is included then the agreement is not as impressive as originally claimed by Eddy. The current consensus is that the alignments are certainly more coincidental than intentional. Another example comes from the ancient Mayas of Central America and the Yucatan, who were exceptionally keen observers of the heavens and built many temples and structures from which to observe the stars and the planets. They were most interested in the planet Venus and constructed a highly accurate calendar based on its heliacal rising. Possible alignments of Mayan temples with Sirius and other stars were investigated in the 1970s; however, with the exception of the temple at Monte Alban, which has a tentative alignment with Sirius, no striking coincidences were found.

Sirius also has an interesting involvement with the ancient constellation of the bow and arrow. The ancient Chinese recognized a constellation of stars forming a bow and an arrow. The bow and arrow resides to the south and east of Sirius and is formed from stars in the modern constellations of Pupis (the stern portion of the old constellation of Argo Navis) and Canis Major. In the Chinese constellation, the arrow is aimed directly at the Heavenly Wolf, Sirius. This is almost directly echoed by the ancient Mesopotamians who appear to have recognized the same constellation, with the exception that Sirius is included as the tip of the arrow. To the Babylonians Sirius was *^{mul} KAK.SI.DI*, "the arrow-star". In later Persian culture, Sirius was known as *Tir* "the Arrow". As mentioned in the first chapter, the sky depicted on the ceiling of the Temple of Hathor at Dendera, Egypt shows the archer–goddess Satet with a drawn bow and an arrow pointed in the direction of Sirius/Hathor. Satet was also the goddess of the hunt, the inundation, and the waters, among other things.

After the ancient Egyptians and their elaborate beliefs surrounding Isis, the Zoroastrian faith, which began in Persia around 600 BC, had perhaps the greatest devotion to Sirius. To the Zoroastrians, *Tishtrya* was the angel of the star Sirius. Their holy book, the *Avesta*, contains a lengthy hymn, the *Tishtar Yasht*, devoted almost entirely to Tishtrya, where many of the verses began with the devotional expression, "We sacrifice unto Tishtrya, the bright and glorious star, who ..." To the Zoroastrians many of the attributes of Sirius are centered on the rains and waters. For example, "We sacrifice unto Tishtrya, the bright and glorious star, for whom long the standing waters, and the running spring-waters, the stream-waters, and the rain-waters ..." Another common theme is that of "a swift-flying and swift-moving arrow that flies to the sea". In ancient India Sirius was often associated with hunting. In Sanskrit its title was the "Deer-Slayer" and the "Hunter". The sacred Hindu texts, the Vedas, also refer to Sirius as *Tishtrya*, and variously as *Tishia*, *Tishiga*, or *Tistar*, "The Chieftain's Star".

In Arabic, Sirius is known as Al Shi'ra, which clearly resembles the Greek and other names of the star. In later Islamic astronomical texts Sirius is called Al Kalb Al Akbar, for the "Greater Dog", following earlier Greek practice. Sirius is mentioned in several places in the Quran, the only star to be explicitly identified, other than the sun. In chapter (surah) 50, called Al Najm (the star), the Quran recounts the attributes of Allah in verse 53 which states, "And he who is the Lord of Sirius". There is also surah

86 which speaks of the night visitor, *At-tariq*, which is the star of piercing brightness, and which some hold to be Sirius:

> 86.1 By the Sky and the Night-Visitant
> 86.2 And what will explain to thee what the Night-Visitant is?
> 86.3 (It is) the Star of piercing brightness.

In Africa, the Dogon tribe, in the modern nation of Mali, is claimed to possess a well-publicized and remarkable set of beliefs concerning Sirius. These beliefs and their possible origin are discussed in more detail in Chapter 11. In addition to these beliefs, the Dogon are said to have noted the heliacal rising of Sirius by the use of stone markers which they used to define the sight lines on the horizon to the points where Sirius will rise just before the sun. They are also said to have made use of a calendar based on Sirius along with solar and lunar calendars, in much the same fashion as the ancient Egyptians. References to these practices go back to the early 1930s, when Dogon customs and beliefs were first studied by the French anthropologist Marcel Griaule. Further mention of these Sirius-related astronomical connections and the "sirien" calendar is contained in the 1965 book *Le Renard pâle* written by Griaule and his colleague, Germaine Dieterlen. In 1998 Dieterlen, accompanied by the French astronomer J. M. Bonnet-Bidaud, revisited some of the sites mentioned by Griaule and conducted measurements of alignments of the markers which seemed to confirm that they could have been used for the purpose of marking the heliacal rising of Sirius.

Since prehistoric times, the skills necessary for navigating between the small island groups and atolls, which spread across the vast expanses of the central Pacific Ocean, required a highly-developed practical knowledge of the stars, the constellations, and their motions. Not surprisingly the ancient Polynesians and Micronesians cultivated many legends and much lore connected with the bright stars of the southern sky, in particular Sirius. These stars served as both navigation beacons and seasonal markers. As navigational aids, the stars were both essential direction finders and latitude indicators for crossing the trackless ocean. When near the horizon, bright stars such as Sirius were used as "star compasses" to define positions on the horizon by which mariners could chart the courses of their outriggers. A succession of such stars, embedded within familiar constellations, were used throughout the night as they rose and set. Navigation using such markers was called following "the star path". The great usefulness of this stellar navigation system was facilitated by the proximity of most of Polynesia to the earth's equator, where stars in the east and west tend to rise and set nearly vertically.

Another key use of certain stars was as latitude indicators. The correct latitude of a particular island or archipelago could be easily found by sailing north or south until the proper star was directly overhead. For example, Sirius sits at a declination of 17°S, which closely matches the latitude of the island of Fiji at 17°S. Thus, Sirius will pass directly overhead on Fiji each night or day. The stars were also seasonal indicators, which determined the best time of year to undertake certain voyages in order to take the best advantage of the seasonal trade winds and currents. In this regard, Sirius was one of the most familiar fixtures of both the southern summer in New Zealand and northern winter skies in the Hawaiian Islands.

One of the most widely recognized of the Polynesian constellations was the *Manu*, or "The Great Bird Constellation". This constellation, with Procyon forming the northern wingtip, Canopus the southern wingtip, and Sirius the body, served to divide the Polynesian sky into two sectors. Sirius was also often associated with the small constellation of the Pleiades. The rising of the Pleiades was the prelude to a train of bright stars: Aldebaran, in Taurus; and Rigel, Bellatrix, and Betelgeuse in Orion; followed by Sirius and Canopus. Sirius was a familiar star throughout Polynesia and was known by many names: to the south in New Zealand it was "Rehua"; in the Marquesas on the equator, it was "Tau-ua"; and in Hawaii to the north it had numerous names such as "Aa" and "Hoku-Kauopae".

Throughout much of Polynesia, Sirius was also a well-known seasonal marker. For example, to the Maori of New Zealand in the south the appearance of Sirius in the morning sky coincided with the freezing cold of winter. Their word "takurua" was synonymous with both Sirius and winter. To the north in Hawaii, Sirius was called "Ka' ulua" and its culmination near midnight signaled the time of a great celebration or festival. Sirius was the "Queen of Heaven" and marked the winter solstice.

Much more could be written about the ancient lore associated with Sirius: the topic is virtually inexhaustible. For those interested in Greek and Roman ideas, in particular those associated with the ancient color of Sirius, Roger Ceragioli's articles in the *Journal for the History of Astronomy* and in *Astronomy and Cultures* are very valuable. Another source is the venerable *Star Names Their Lore and Meaning* by Richard Allen, which contains a potpourri of names and associations. For some of the more primal and enigmatic links to Sirius, there is *Hamlet's Mill* by Giorgio de Santillana and Hertha von Dechend. Finally, a very contemporary mythological link with Sirius is the belief system of the Dogon, a sub-Saharan tribe in the modern African nation of Mali. Again, this controversial subject is covered in detail in Chapter 11.

Part Two

The Nature of the Stars

What are the stars? How does their nature differ from the five "wandering stars" or planets known since antiquity, and from the sun and the moon? How is the universe of stars and planets arranged and how does it function? Over the ages there have been many answers to these questions: superstitious, mythological, religious, philosophical, and ultimately scientific. The long process of acquiring useful and reliable knowledge about the stars has been punctuated with a number of discoveries and insights that are important to the story of Sirius, and in which Sirius often played a critical role.

3

From Myth to Reality

"That the sun and stars were fiery or red-hot Stones and Golden Clods"
James Gregory, 1702,
paraphrasing Diogenes Laertius on the life of Anaxagorus

The long intellectual journey from viewing Sirius and the other stars as the heavenly counterparts of myth and superstition, to the realization that the stars are actual material objects governed by physical laws, spanned nearly 25 centuries. One important way station on that journey was the ancient city of Alexandria in Egypt. It was in Alexandria, founded by Alexander the Great in 332 BC during his conquest of Egypt, where many key Greek astronomers either worked or studied over a period of nearly five centuries. The origin of many important discoveries and ideas that shape our thoughts about the stars can be traced to Alexandria and to the Great Library that flourished there. Among the important names associated with Alexandrian astronomy are Aristarchus, Hipparchus, Eratosthenes, and Ptolemy. With the prominent exception of Ptolemy, much of the original work of these ancient astronomers is now lost and we have only secondary sources and commentaries from which to gauge their achievements.

It was Aristarchus of Samos (c. 310 BC–c. 230 BC) who is credited with making the first practical attempts, based on geometry, to estimate the actual sizes of the sun and the moon and their distances from the earth. Although he underestimated the true distances by a large factor, his methods were sound. Aristarchus is also credited with first advocating a heliocentric model of the universe, in which the earth and the planets orbited the sun and the daily motion of the heavens was due to the rotation of the earth on its axis. Unfortunately, other than his treatise on the estimated distances of the sun and the moon, no complete works from Aristarchus remain. However, at least one ancient source reports that along with his heliocentric theory, Aristarchus placed the stars at very large distances and regarded them as distant suns. He needed to remove the stars to great distances in order to explain why the earth's annual orbit of the sun did not result in noticeable angular shifts in the positions of the stars, as they were viewed first from one side of the earth's orbit and then the other. It is this lack of a measurable parallax shift (see Appendix A) among the stars that was to plague his, and later, heliocentric theories. Regrettably, for astronomy Aristarchus' early heliocentric ideas attracted few followers. His heliocentric model was criticized (which is one reason we know about it at all) by the mathematician Archimedes, who focused on an illogical proportion that Aristarchus used, which seemed to imply an infinite distance for the stars. Aristarchus was followed by Eratosthenes of Cyrene (c. 274 BC–190 BC), who first estimated the size of the earth, arriving at a value quite close to its actual size. Eratosthenes also made very credible estimates of the distance of the sun and moon and the tilt of the earth's equatorial plane to the plane of the ecliptic.

Hipparchus of Rhodes (c. 190 BC–120 BC) is credited with discovering the important motion of the stars, called precession. Precession (see Appendix A) is characterized by a slow motion of the direction of the earth's rotational pole with respect to the stars. The origin of this motion is the gravitational force of the moon and the sun on the earth's equatorial bulge, which causes the pole to slowly trace out a conical path, of nearly constant angle, with respect to the earth's orbital plane. One of the manifestations of precession is that the line defined by the intersection of the earth's equatorial plane with its orbital plane, slowly moves through the stars of the zodiac. It is an almost imperceptible motion of about 50.3 arc seconds per century, but over time it causes the equinoxes to shift with respect to the stars and to eventually result in a succession of different pole stars. For example, the present celestial pole is located near the star Polaris in the constellation Ursa Minor. But for the ancient Egyptians 4000 years ago the pole was located near the star Thuban in the constellation of Draco. Since most ancients thought the earth was stationary, precession like the diurnal motion of the heavens had to be accommodated by assigning these motions to the stars themselves.

The Roman author Pliny the Elder, writing in the 1st century AD, mentions that Hipparchus noticed a "nova", or new star, appear in the constellation of Scorpio. This event prompted Hipparchus to begin carefully recording measurements of stellar positions to determine if stars might occasionally appear and disappear, or if they might move with respect to one another. Hipparchus compiled these data into a catalog containing perhaps 850 stars sometime around 129 BC. Although this catalog no longer exists, it was used by later Greek astronomers as the basis for subsequent star catalogs. Hipparchus' star catalog appears to be the source of the positions of the constellations as depicted on the famous Farnese Atlas, a Roman copy of an earlier Greek statue showing the giant Atlas supporting a marble globe on which 41 of the original 48 Greek constellations are carved. We know this because in 2004 Dr. Bradley Schaefer of the University of Louisiana, at Baton Rouge, carefully measured the locations of the constellations relative to the latitudinal and meridional lines on the globe. Schaefer concluded that due to the changes caused by precession, the sky portrayed on the stone globe corresponds to an historical date of 125 BC, plus or minus 55 years, and to observations that were made at approximately the latitude of

Figure 3.1. An imagined likeness of Claudius Ptolemy (*Great Astronomers*, R. S. Ball, 1907).

the island of Rhodes, which lies about 5° of latitude north of Alexandria. Hipparchus is known to have worked at Rhodes, and it is Schaefer's conclusion that the stellar positions used to fashion the globe were taken from Hipparchus' lost star catalog.

Claudius Ptolemy (85 AD–165 AD, Figure 3.1) was arguably the most influential of all ancient astronomers. His name indicates he was a Roman citizen of Greek descent who worked and studied in Alexandria. Unlike the works of many of those who came before him, later Arabic and Latin translations of several of Ptolemy's complete books still exist. The most important of these is his *Almagest*, which was probably written around 145 AD. The name *Almagest* comes from a poor rendering of the Arabic title for the book, *Al Majisti*, or "The Greatest". The original Greek title was *Syntaxis* or "The Compilation". This ancient textbook contains virtually all that was known of the heavens and the stars during classical times. Most importantly the *Almagest* described the first working mathematical model of the apparent motions of the planets. Earlier, Greeks had developed a system based on a compound series of uniform circular motions to describe planetary motion; however, it was the innovations of Ptolemy which forged these concepts into practical descriptions of the complex movements of the planets, the sun, and the moon.

According to the Ptolemaic system (see Appendix A), the universe consisted of a stationary spherical earth surrounded by a series of roughly concentric transparent crystal spheres. The rotation of these spheres, one for each planetary body and one containing all of the stars, generated the observed motion of the heavens. The complexity of these motions was produced by a sequence of basic uniform and circular motions. In order to accommodate the obvious nonuniform and noncircular motions of the planets, it was therefore necessary to postulate a set of multiple compound circular motions for each planet. Ptolemy, building on older Greek ideas, first had each planet circle the earth on a large circle called a deferent. Attached to each deferent was a smaller circle called an epicycle, which rotated as it traveled around the deferent. Ptolemy's critical addition to this scheme was to offset the center of each deferent from the earth. Then on the other side of the center of motion was placed the equant, an empty point in space equidistant from the center and opposite the earth. The function of the equant was to regulate the motion of the epicycle. Ptolemy's invention of the equant represented a small but significant "fudge" to the concept of uniform circular motion. It was from the perspective of the imaginary equant, not the earth, that the epicycle exhibited uniform motion. By carefully defining the size of the deferents, epicycles, and equants of each planet, and assigning the necessary angular rotation rates, Ptolemy could reproduce the apparent angular motions of the planets from a geocentric point of view, with a tolerable accuracy. Ptolemy's model maintained a reasonably good representation of the skies over periods of centuries and was widely used for the next 1500 years. It is also one of the first workable mathematical descriptions of the detailed motions of the then observable universe. In Ptolemy's system there were seven planetary spheres, consisting of one deferent and one epicycle for each of the five known planets plus spheres for the sun and the moon, which were considered planets at the time. The stars in contrast were confined to fixed locations on an eighth sphere, which completed a singular uniform revolution of the earth each day. Beyond the sphere of the stars was a ninth sphere that imparted motion to the other spheres. A tenth sphere was necessary to accommodate the precessional motion of the stars.

In books VII and VIII of the *Almagest* Ptolemy deals with the stars. He began by first establishing that the stars are indeed fixed, by examining a number of linear alignments of three or more stars and showing that there had been no relative change in their positions in the 260 years since the time of Hipparchus. One implication of this discovery was that any motion of the stars, diurnal rotation or precession, must be a shared motion of all stars. Ptolemy also describes the construction of a globe on which to depict the location of the stars. The globe was to be dark and contain the celestial equator marked off in 360 degrees, and on the globe the stars were to be painted in yellow and sized according to their apparent brightness. Six of the brighter stars, including Sirius, were to be painted red. A movable ecliptic equator was also described. Ptolemy chose Sirius to define the location of the globe's central meridian.

A major portion of the *Almagest*'s book VII is occupied by a catalog listing, by constellation, 1022 stars which can be seen with the naked eye from the latitude of Alexandria. This list is presented in tabular form giving the ecliptic coordinates (see Appendix A) and a brief description of the relative location for each star within its constellation. For example, the star Arcturus in the constellation Boötes is "the star between the thighs called 'Arcturus', reddish". Also provided are the magnitudes of each star. Ptolemy ranks the brightest stars such as Sirius as 1st magnitude and the faintest stars as 6th magnitude. In this context the term magnitude is clearly associated with the concept of "size" or "importance". He mentions the "color" of only six stars. Five, such as Betelgeuse and Arcturus, are correctly described as "reddish". Curiously, Sirius is among the "reddish" stars: *The star is, on the mouth, brightest, called the "Dog" and reddish.* Given that Sirius is not red at all, but a distinct blue–white star, this curious description has inspired a seemingly endless debate as to why Ptolemy used the term reddish in connection with Sirius (see Chapter 10).

It is unclear how much of Ptolemy's star catalog consisted of observations he had made himself and how much represented borrowed updates from earlier catalogs such as that of Hipparchus. What is not in doubt, however, is that Ptolemy's description of the sky persisted and was only superseded in Europe by the star catalog of the Danish astronomer Tycho Brahe in 1600 AD. Ptolemy said very little concerning the physical nature of the stars and what they were made of. For the most part, the answers to such questions were to be found in the philosophy and physics of Aristotle (384 BC–322 BC). According to Aristotle, the substance of the stars, like that of the planets and celestial spheres, was said to be the celestial ether, which was changeless and possessed a naturally circular state of motion.

Although there were other schools of Greek philosophy which rejected the notions of Aristotle, it was the cosmos described by the motions of Ptolemy and physics of Aristotle that dominated astronomy for nearly 1500 years. In the West the working knowledge of this cosmos was slowly forgotten after the fall of Rome. In the East it persisted as a philosophical system, but was shunned by the Byzantine church. It was the Moslems, beginning in the 10th century, who revived these ideas, finding them useful for predicting the positions of the sun, the moon, and the planets as well as furnishing a satisfyingly abstract and orderly way to view the cosmos. Islamic scholars translated many Greek texts into Arabic, including Ptolemy's *Syntaxis* (the *Almagest*) and studied them in centers of higher learning from Samarkand in Central Asia to Baghdad in the Middle East, and to Granada in Spain in the west. Although Islamic scholars made improvements in Ptolemy's model and some thinkers even expressed doubts about the reality and nature of the crystal spheres, the basic classical scheme and interpretation of the cosmos was retained. One high point of Islamic astronomy occurred when Ulugh Begh, the grandson of the Mongol conqueror Tamerlane, built in 1427 a large observatory dedicated to astronomy in the ancient Silk Road city of Samarkand (in modern day Uzbekistan). There his astronomers reobserved the positions and magnitudes of the stars contained in Ptolemy's catalog. This catalog, called the *Zij-i-Sultani*, represented a notable improvement over Ptolemy's catalog.

It was from Islamic sources, first from Moslem Spain around 900 AD and later from the crumbling Byzantine Empire in the east, that first Arabic and later Greek texts entered Medieval Europe. This reintroduction of classical astronomical learning into Europe provoked a slow but steady adoption of new ideas regarding the stars and the universe.

The view of the stars that dominated the thought of late Medieval Europe was embedded in the Ptolemaic cosmological view. To the scholastics and theologians that thought about and debated such matters, the universe was still described by the physics and metaphysics of Aristotle and the planetary mechanics of Ptolemy, but it was made compatible with the theology of the Christian Church. At the center of this finite and spherical universe, rested the immobile earth, where the four elements—earth, water, air, and fire—were in constant flux. Upon the earth and up to the level of the first celestial sphere, defined by the moon, *change* was dominant and pervasive. Birth, growth, death, and decay proceeded as they always had. Heavy base material, such as stones, fell towards the center of the earth and light ethereal bodies, such as flames, ascended. Beyond the sphere of the moon lay an entirely different realm. Here the planets, including the principal planet, the sun, rotated in silent eternal majesty, conveyed by an immense hierarchy of invisible crystal spheres. In this extra-lunar realm, by definition, no change was possible and the substance of this level of the universe was the fifth element, the celestial ether.

Although the earth occupied the physical center of this universe, it did not occupy its spiritual focus, but rather occupied the very bottom level of an ascending hierarchy. Above the earth rotated the celestial sphere of the moon followed by those of Mercury and Venus. Next came the sun, regarded as the most important of the "planets". External to the solar sphere were the spheres of Mars, Jupiter, and Saturn. Beyond Saturn was the eighth celestial sphere of the stars. This entire set of nested transparent spheres was driven by a mechanism residing in the ninth outer sphere, the *primum mobile* or prime mover, as required by Aristotelian physics. Sometimes a tenth sphere was added to account for the precessional motion of the stars. A late Christian addition to this heavenly construct was a region of celestial light and the abode of god and the angels, known as the Empyrean. It was the Empyrean where the "music of the spheres" was believed to be generated and where Dante was led by Beatrice in the *Divine Comedy*.

The working details of this universe were a matter of debate and speculation. Was the number of stars countless, or simply the 1022 catalogued by Ptolemy? Were the stars self-luminous or did they reflect the light from the sun? How distant was the stellar sphere and did it possess a substantial thickness? Although there were differing opinions on these points, the consensus was that Ptolemy's visible stars populated a thin eighth sphere, and like the planets they merely reflected the light of the sun. A common estimate was the stars resided at a distance of 73 million miles, in modern terms this would only be about three-quarters the distance of the earth to the sun. The stars, like the planets, were believed to be the result of dense accumulations of Aristotle's all pervasive celestial ether.

The person who first successfully challenged this medieval view of the universe and the stars was Nicholas Copernicus (1473–1543), a Church lawyer working in the small town of Frombrok on the Baltic coast of northern Poland. In 1543 Copernicus published his book *De revolutionibus* which reordered the solar system by placing the sun at the center and having the earth spin on its axis and orbit the sun along with the other planets. Three decades earlier, Copernicus had circulated a limited number of manuscripts which briefly outlined his revolutionary ideas. With *De revolutionibus*, however, he provided the details and argued the case for a moving earth. Copernicus, who died in the year that *De revolutionibus* first appeared, had dedicated the book to Pope Paul III, probably as a means of deflecting any ecclesiastical criticism.

Copernicus' universe had a compelling simplicity that the Ptolemaic system lacked, while at the same time producing predictions of planetary motion that were on a par with the Ptolemaic system. Copernicus, however, retained the uniform circular motions and epicycles (but not the equants) that were part of the old Ptolemaic system.

The initial reaction to *De revolutionibus* was muted by an unauthorized introduction penned by the book's Nuremberg editor, which argued that the physical model offered by Copernicus need not necessarily be true for the mathematical descriptions to be valid. Indeed, many early readers adopted this abstracted view of the heavens, and for decades both the Ptolemaic and Copernican models were taught in some European universities. It was only seven decades later, when some scholars began to insist on the physical reality of the Copernican system, that the Catholic Church took steps to proscribe *De revolutionibus*, "subject to correction". Even then, it was the motion of the earth and not its centrality that was the chief issue.

Most discussions of Copernicus rightfully focus on his removal of the earth from the center of the universe and of planetary motion, and its replacement by the sun. However, Copernicus also made several revolutionary changes with respect to the stars. First and foremost he stopped the stars from madly whirling overhead on a daily basis and set them at rest. He not only halted the stars but he established them, along with the sun, as the fixed and immobile components of the universe. In fact Copernicus even relieved the stars of their slow precessional motion. He did this by tying the origin of the ecliptic coordinate system (Appendix A) to the stars. He picked the star Mesarthim, in the western horn of the constellation of Aries the Ram, as the origin of the ecliptic longitude system. He then recalculated the ecliptic longitudes of all the stars in Ptolemy's catalog, correcting for 1422 years of precession. Precession and diurnal rotation were now effectively earthly, not stellar motions. Copernicus also placed the stars at a vast unspecified distance. He did this to avoid the obvious problem that the stars exhibited no annual parallax shifts, as should be expected with his moving earth. This lack of a measurable parallax remained a powerful counter-argument to Copernican ideas well into the 17th century.

While the debate regarding the ideas of Copernicus centered on the motion of the earth, two early advocates of the Copernican model contributed what would prove to be two very fruitful new ideas regarding the stars. The first idea appeared in 1576 in a short work by the Englishman Thomas Digges (1546–1595) entitled *A Perfit Description of the Caelestiall Orbes*, summarizing the Copernican model. When Digges got to the stars, he completely departed from Copernicus, and added his own unique interpretation of the stellar universe. There was no eighth sphere: the stars were scattered throughout a limitless volume centered on the sun. Moreover, the explanation for the stars having different brightnesses lay in their differing distances, not in their different sizes. Further, the sky contained hosts of stars too faint to see with the human eye, "lightes innumerable" and "without ende", he called them. Digges accompanied his description with a now famous diagram making clear how he viewed the stellar universe (Figure 3.2).

How did Digges come by this rather startling insight? It appears that his father Leonard Digges, a military engineer, had conducted experiments with what was called

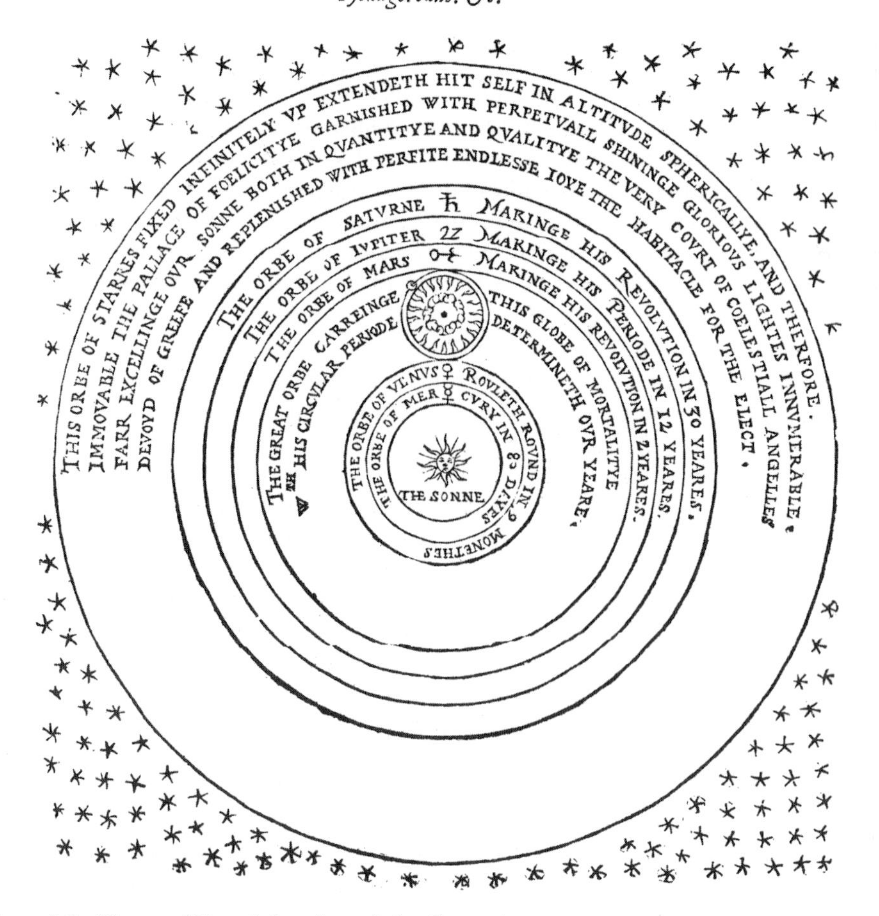

Figure 3.2. Thomas Digges' drawing of the Copernican System and Digges' concept of the uniform distribution of the stars (from *A Prognostication Everlasting*, 1576, permission, The Huntington Library).

a "perspective glass", an early pre-Galilean telescope consisting of a convex objective lens and concave mirror. It is not hard to imagine that among the things viewed with this "perspective glass" was the sky containing a host of previously unseen stars. Digges and his ideas, including his endless realm of the stars, apparently had an impact on a number of prominent citizens of Elizabethan England.

The second big idea came not from an astronomer but from a former Dominican monk, turned philosopher and heretic, Giordano Bruno (1548–1600). Bruno had fled his native Italy in 1576 after refusing to renounce a number of heretical views. He spent the next decade in exile wandering throughout Protestant Europe. Beginning in

1583, he resided for two years in England where he met with a number of prominent English intellectuals. In 1584, while in London, Bruno published a series of books in vernacular Italian which contained, among a great many other things, his ideas about the stars. According to Bruno the stars are but distant suns arrayed in an infinite universe having no center. Starting from the idea that the sun was a star he reasoned that, like the sun, the stars ought also to be orbited by planets and that these planets would be inhabited. Bruno was a strong advocate of Copernicus but he had no astronomical training and seems rather to have arrived at his insights about the stars though metaphysical arguments. Although there is no proof, he may have well become aware of the ideas of Digges during his stay in England. Bruno returned to Italy in 1591, and in 1599 he was convicted of heresy and burned at the stake on February 17, 1600, for refusing to recant his ideas. Although his views of the stars would have been considered heretical at the time, it is likely that it was primarily Bruno's many strong anti-clerical views that led to his execution.

Early in the evening of November 11, 1572, an intense, previously unknown star suddenly appeared in the Constellation of Cassiopeia. Among the few Europeans who noticed this singular event was a young Danish nobleman with an interest in astronomy, Tycho Brahe (1546–1601). Over the next six months, before the star's light dulled and faded, he made repeated measurements of its position, brightness, and appearance. Tycho determined that the star did not move, as a comet would nor did it have a comet's "hairy" appearance. After repeated measurements to determine its diurnal parallax, he concluded that the star must lie beyond the orbit of the moon and among the stars of the eighth sphere. Tycho published a book, *De Stella Nova* or "of the New Star" describing his observations and conclusions. The sudden appearance of this star, followed by its slow fading, convinced Tycho that the old Ptolemaic universe was false and that the ethereal realms of Aristotle and Ptolemy were not unchangeable. It was also this new star that thrust Tycho into a lifelong career as an astronomer. He later went on to construct a well-staffed private observatory on the tiny island of Hven in the strait between Denmark and Sweden. There for the next twenty years he made the most precise observations of the positions of the planets and the stars, in the era before the invention of the telescope. Interestingly, one of the few Europeans who also witnessed and wrote of the new star was Thomas Digges, in England.

Tycho, who already doubted Ptolemy's model, was unable to detect the parallax of any star and this led him to also doubt Copernicus' moving earth. His solution was to develop his own hybrid model of the universe with a central, rotating, but otherwise stationary, earth. In this geocentric model, the sun and the moon orbited the earth but all the other planets orbited the sun. The stars remained in a distant far off shell. One of the motivations for Tycho's observations was to gather precise measurements of the positions of the planets, in support of his "Tychonic System". A necessary prerequisite for accurate planetary measurements was a precise star catalog. What finally emerged in 1598 was a catalog of some 1004 stars, all measured visually, with an unheard-of level of precision of nearly one arc minute. To mathematically reduce his voluminous observations, Tycho hired assistants. Among those who eventually assisted Tycho during his last years in Prague, was a young German schoolteacher and talented mathematician named Johannes Kepler (1571–1630).

Figure 3.3. Johannes Kepler (*Great Astronomers*, R. S. Ball, 1907).

Just as he was completing his catalog, Tycho ran into difficulties with the Royal Danish Court and was effectively exiled from Denmark and his island observatory. He and his entourage finally settled in Prague where Tycho died in 1601. It was Kepler (Figure 3.3) who inherited Tycho's observations, along with Tycho's position as court astronomer to Rudolph II, the Holy Roman Emperor. Tycho's dying hope was that Kepler, the clever mathematician that he had rescued from obscurity, would use Tycho's observations to demonstrate the reality of the Tychonic System. Kepler, however, remained a confirmed Copernican and secretly doubted his benefactor's Tychonic System.

In 1604 Kepler was deeply involved in an effort to reconcile the theoretical motions of Mars with Tycho's precise observations. On a cold October morning of that year Kepler was awakened with news of a brilliant new star in the constellation of Taurus. Realizing the importance of this new star, Kepler dropped everything and devoted himself to its intense study. Like Tycho, Kepler also published a book on this remarkable phenomenon entitled *De Stella Nova*. Kepler, as Tycho had, also showed that the new star had no perceptible motion and must lie among other stars. The appearance of two brilliant new stars, in the space of just thirty years, did much to demolish Aristotle's doctrine of a spherical domain of unchangeable stars, and helped to undermine the reality of the crystal spheres and the artificial mechanics of the Ptolemaic model. Nevertheless, in the minds of the two chroniclers, Tycho and Kepler and many others, these two events did not entirely remove the sphere of the stars. Tycho, who held to his earth-centered Tychonic System and Kepler, who viewed the sun as the central source of luminosity in the universe, regarded the stars as affixed to a distant sphere, which bounded the knowable universe.

The year 1604 also saw the publication of Kepler's book *Astronomiae Pars Optica*, in which he made a seemingly simple observation that would turn out to be an important and fundamental discovery about the nature of light. It is commonsense that a source of light, such as a lantern in the night, will appear to diminish in brightness as it is viewed from ever greater distances. Kepler was the first person to answer the question of exactly how the brightness appears to change with distance. The light will diminish inversely as the square of the distance. A distant lamp will thus appear one-quarter as bright when it is twice as far away. Viewed another way, this simple law provides a way to determine the relative distances of the stars from the earth by their relative apparent brightnesses, *provided all stars have a similar luminosity*. In the latter half of the 17th century, it would be the ingenious use of this principle which would lead to the first realistic estimates of the truly immense distances to the stars.

Kepler is best known for his three laws of planetary motion (see Appendix A). He arrived at the first two laws in 1609 only after many intense and frustrating attempts to mathematically reconcile Tycho's accurate positions of Mars with various combinations of compound circular motions. He finally achieved success after completely abandoning the idea of circular motions and considering other possibilities, such as that Mars might move in an elliptical orbit. Kepler's revolutionary first law states that the planets move in elliptical orbits, with the sun at one of the foci. It eliminated the foundation of the Ptolemaic, Copernican, and Tychonic models, which assumed circular and compound circular paths. Kepler's second law states that a planet sweeps out equal areas in equal times. This eliminated the universal assumption of uniform motions. The planets actually speed up when near the sun and slow down when farther away, in such a way that a line from the sun to the planet will sweep out an area at a constant rate. These two simply stated laws not only broke the ancient mold of uniform circular motion but yielded an immense increase in the precision of the determination of planetary orbits.

Kepler arrived at his third law in 1618. This law relates the orbital period of a planet to its average distance from the sun and states that the square of the period is equal to the cube of the distance ($P^2 \propto d^3$). The earth has an orbital period of one year and its average distance from the sun is 149 million km, or one astronomical unit (AU). Using these units of time and distance based on the motion of the earth, Jupiter, according to Kepler's third law, has a period of 11.863 years and a distance of 5.202 AU from the sun. Together Kepler's three laws made it clear that it was the sun that was the center of motion in the solar system. These laws, however, were completely empirical and shed little light on the nature of the motive force emanating from the sun. Additionally Kepler's laws seemed to apply only to the solar system and have no obvious application to the stars.

From his backyard garden in Padua, Galileo (1564–1642) first turned his new telescope to the heavens in the winter of 1609–10. There he saw not only the craters and mountains of the moon and the moons of Jupiter, he also observed a myriad of stars. He promptly reported these discoveries in his small book *Sidereus Nuncius* or "A Message from the Stars", which was published in March 1610. Although history focuses on the impact that his observations had on the foundations of the Ptolemaic order, his description of the stars was also important. Galileo was astonished how the familiar 6 or 7 stars that can be seen by eye in the small constellation of the Pleiades, were transformed into a field containing 43 stars when viewed with his telescope. He described how his telescope resolved the faint glow of the Milky Way into clouds of innumerable individual stars. He also marveled at how his telescope increased the apparent brightness of faint 5th and 6th magnitude stars into stars which appeared as bright as Sirius. However, the stars, unlike the planets, stubbornly remained unresolved points of light in his telescope.

Galileo chose not to speculate on the meaning of the thousands of previously unseen stars revealed by the telescope. He was surely aware of the assertions of Bruno concerning a limitless universe of stars, but he was also well aware of Bruno's fate. Moreover, Galileo was an eminently practical scientist and not well disposed to the sort of unsupported philosophical arguments that characterized Bruno's ideas concerning the stars. Nevertheless, Galileo had witnessed and even lectured on the nova of 1604, and became a Copernican well before he built his first telescope. Additionally, the experiments he conducted on motion and on falling bodies had also caused him to doubt Aristotelian physics. Galileo was fully cognizant of the power of the idea that the stars were distant suns and freely used this concept later in his career.

Galileo directly entered the debate between the Copernicans and the adherents of Ptolemy in 1632, with a book that ultimately provoked the infamous reaction by the Catholic Church that led to Galileo's subsequent heresy trial. Entitled "Dialog on the Two Chief World Systems: Ptolemaic and Copernican", or *Dialogo*, Galileo used the popular literary device of a fictitious set of conversations between three individuals: Salviati, an eloquent and wise Copernican; Simplicio, a less than convincing advocate of the Ptolemaic universe; and a third person, Sagredo, who was ostensibly neutral and mediated the dialog by offering wise judgments on the arguments being made. One of the critical objections raised by Simplicio concerned the distance to the stars and the lack of an observable stellar parallax. In Galileo's day, the predominate view was that the stars were relatively nearby and possessed relatively large angular diameters of just under two arc minutes. Both Tycho and Kepler held this view. In *Dialogo*, Galileo, speaking through Salviati, refutes this belief by describing a simple experiment that Galileo actually performed with the bright star Vega in the constellation of Lyra. He hung a thin vertical thread and then retreated to a position where the thread just obscured the light from Vega, when viewed with one eye. From the distance between his eye and the thread and the width of the thread, Galileo estimated that Vega must have an angular diameter smaller than 5 arc seconds (an angle of arc second is 1/3600 of one degree); or about 25 times smaller than the common estimates of his day. Using this number and assuming that Vega had a diameter similar to that of the sun, Galileo then estimated that Vega had to lie at a distance at least 360 times farther than the orbit of Saturn.

Later in *Dialogo* Galileo described two methods for measuring the small parallaxes of distant stars. The first involved fixing a telescope so that it always pointed in a constant direction and watching for the annual parallax motion of a star that passed nightly through the telescope's field of view. The second method made implicit use of the idea that the stars are distant suns. In this method close pairs of stars would be sought out with the telescope, in which one star was very much brighter than the second. Galileo, assuming all stars had similar luminosities, reasoned that the fainter star would therefore be much more distant than the brighter star and thus have a much smaller relative parallax compared with that of the brighter. The small relative angular shift between the two stars could then be readily discerned with the telescope. Although Galileo never attempted either method, both were eventually employed in attempts to detect stellar parallax shifts. What was ultimately discovered was not parallax, but two unexpected aspects of the stars. The first method, of using a fixed telescope to monitor the changes in the annual position of the stars, led to the discovery of a new kind of "stellar motion", called the aberration of starlight. This motion results from the finite speed of light which produces an apparent shift in a star's position due to the earth's orbital motion (see Appendix A). It was stellar aberration, not stellar parallax which produced the first demonstrable evidence of the earth's motion. The second method of looking for small relative parallax shifts between close stellar pairs, of different brightness, yielded another discovery which would ultimately undermine the idea that all stars had the same luminosity as the sun.

The name Edmund Halley (1656–1742) is forever linked to Halley's Comet, whose return he predicted late in life. This popular association, however, obscures Halley's long and illustrious career, which consisted of far more than the study of cometary orbits. In 1676, as a 21-year-old student, fresh from Oxford University, Halley had conceived the bold idea of undertaking a dangerous sea voyage to map the stars of the southern hemisphere. He spent an entire year on the isolated island of St. Helena in the middle of the south Atlantic, struggling under the very difficult conditions, to produce the first accurate star catalog of the southern skies. Forty years later, in 1717, while studying the precession of the equinoxes, Halley made a most astonishing discovery. In comparing the contemporary north–south positions of stars with those given by Ptolemy's *Almagest*, Halley discovered that the bright stars, Sirius and Arcturus had measurably changed their positions since antiquity. In the case of Sirius, it had moved 30 arc minutes to the south. In fact, Sirius is one of the more rapidly moving stars, drifting to the southwest at a rate of 1.34 arc seconds per year; or covering the distance corresponding to the diameter of the moon in about 1800 years. Halley, implicitly using the idea that the stars are distant suns, recognized that the nearest and brightest stars ought to be those with the largest motions.

Halley had discovered that the stars were in reality not fixed at all, but had slow apparent motions which are now called proper motions. Philosophically this, once and for all, destroyed the notion of the firmament of heaven containing stationary stars. The stars, and by implication the sun, were adrift with respect to one another in a vast void. An immediate practical consequence of this discovery was that, in order to accurately map the heavens, not only the positions of the stars would need to be determined but their motions would now also have to be measured. In practice, this meant that the positions of the stars had to be accurately determined for a minimum of two epochs, separated by a substantial period of time on the order of decades to perhaps a century. In Chapter 4 we will see how one effort to measure stellar proper motions led to another surprising discovery regarding the star Sirius.

In the summer of 1686 a book appeared that fundamentally changed our view of the universe and also the stars. The book was titled *Philosophiae Naturalis Principia Mathematica*, or *Principia* for short. Its author was a reclusive, but well-known Cambridge professor, Isaac Newton (1642–1727). The 45-year-old Newton had produced the book in an astonishingly brief period of just 18 months of intensive writing and research. He had in his notebooks the outlines of some of his ideas, but he produced the bulk of his remarkable results as he wrote. The genesis of the book seems to have been a question posed to Newton during a visit by Edmund Halley in the summer of 1684: What would be the mathematical description of the path of a body moving under the influence of an inverse square law of attraction? What Halley had in mind was, assuming that the sun attracted the planets with a centrally directed force diminishing with the square of the distance, what would be the shape of the corresponding orbits? Without hesitation Newton answered: it would be "an ellipse". Halley then asked how he knew this. "I have calculated it," said Newton. The answer astonished Halley because although the inverse square law had already been guessed by several people (including Halley) from the nature of Kepler's third law, no one could show that an inverse square law must lead to an elliptical orbit, as stated by Kepler's first law.

Several months after Halley's encounter with Newton, he received the draft of a short paper that demonstrated how each of Kepler's three laws mathematically followed from an attractive centrally directed inverse square law. At this point Newton was already deeply involved in the production of *Principia*. *Principia* was an intentionally difficult book written in Latin with convoluted geometrically based derivations of numerous propositions and theorems. Nevertheless, for those who were able to penetrate it, the book contained the long-sought answers to numerous questions. Among these questions were why the earth, the sun and the planets had spherical shapes, a physical explanation of the ocean tides, and the revelation of the cause of the precessional motion of the stars. Newton also showed that the earth's rotation would produce an equatorial bulge in its shape and calculated the size of this bulge. The bulge (of some 20 km) he showed was responsible for the precession of the equinoxes. The gravitational force of the moon and the sun on this equatorial bulge caused the axis of the earth to precess like the motion of a spinning top, which is not aligned with the vertical.

The core of *Principia* was Newton's three laws of motion and his law of gravity (see Appendix A). From these laws it was possible to build a precise mechanics of the solar system, which allowed the calculation of the motion of the planets, including the mutual gravitational influences of the planets on each other. It also explained the orbits of the comets as well as the moons that were observed to orbit Jupiter and

Saturn. *Principia* even contained the tools for calculating an accurate orbit of the moon, but that problem proved so complex that it kept astronomers and mathematicians occupied for almost three centuries. At the time when *Principia* appeared, distances within the solar system were not well known and were measured in units of the earth's distance from the sun, the astronomical unit. Likewise the masses of the planets were not known but they could be measured with respect to the mass of the sun in units of the solar mass. For example, the mass ratios of Jupiter and Saturn with respect to the sun could be measured from observations of the orbits of their moons. It was only after solar system distances were accurately measured and the constant which defines the strength of the gravitational force was measured by Henry Cavendish in 1798, that distances and masses could be usefully measured in terms of terrestrial units such as kilometers and kilograms.

When Isaac Newton published the first edition of his *Principia* in 1686 he more or less neglected the stars. As far as Newton was concerned the stars were distant and effectively fixed, and they served primarily as a convenient reference system for the measurement of planetary motion. However, during preparations for the second edition of *Principia*, Newton, at the behest of several members of the clergy, began to consider the dynamical implications of a universe composed of a uniform distribution of stars. He realized that if the universe of stars was finite then it would eventually collapse, due to mutual gravitational forces. To prevent such a collapse, two solutions seemed possible. The first was an infinite stellar universe in which the net gravitational forces were effectively self-canceling. The second was a finite universe in which the position of each star was exquisitely balanced, so that a gravitational collapse would be prevented. The first possibility implied that the aggregate light from ever more distant stars increased more rapidly than the distance diminished the light of these stars. This led to the dilemma that an infinite universe of stars would produce a sky flooded with a uniform blinding brilliance. The second possibility, which Newton favored, seemed to require a purposeful arrangement of the stars and an active intelligent intervention to forestall gravitational collapse. Divine Providence, in Newton's mind, seemed to be required to sustain such a stellar universe. One obvious way to minimize the necessity for such active intervention was to place the stars at very great distances where the mutual gravitational forces, which diminish as the square of the distance, are minimized. Newton attempted to estimate how distant the stars might actually be. His work on this problem never made it into the second edition of *Principia*, but was contained in his notes and a posthumous book called *The System of the World* published in 1728.

Newton had actually first investigated the distances to the stars in 1685. Realizing the difficulty of measuring stellar parallaxes, he improved upon a technique first suggested by James Gregory, a Scottish scientist. Gregory (1638–1675) had revived the idea of the stars being distant suns and took it one step further in 1668; by adding the assumption that the stars were also identical in all respects to the sun, including brightness. Therefore, to estimate stellar distances it was only necessary to ask how far the sun would have to be removed in order to appear as bright as the brightest stars. Gregory, using Kepler's principle that the intensity of light diminishes with the inverse square of the distance, actually made such a determination by comparing the brightness of Sirius and the planet Jupiter. Unfortunately the answer he obtained was a woeful underestimate, due in part to an inaccurate value of the astronomical unit, the distance between the sun and the earth. Newton, who had access to a better value of the astronomical unit, used the planet Saturn to estimate the distance to a typical first magnitude star, such as Sirius. The answer he got was that a 1st-magnitude star lay at a distance of 1 million times that of Saturn. This value is not too far off the modern distance of 600,000 times the distance of Saturn. Newton, and others who followed him in using this method, were unaware that the identical luminosity of all stars is a very poor approximation. For example, in visual light, Sirius is some 23 times more luminous than the sun. Nevertheless, these early photometric methods served to demonstrate the vast distances to the stars.

Around 10 p.m. on the evening of Tuesday, March 13, 1781, a musician and amateur astronomer was scanning the skies of Bath, England with a home-made reflecting telescope. As he examined a field of stars in the constellation of Gemini, he noticed a relatively bright object with a distinct nonstellar appearance. Thinking he had discovered a new comet, he followed it for the next several weeks as it slowly moved among the stars of the ecliptic. Although it did not resemble or move as a comet, he submitted a paper entitled *An Account of a Comet* to the Royal Society in London in late April. It required several months before astronomers determined an orbit for this new object and realized that it was no "comet" but a new planet beyond the orbit of Saturn. The amateur astronomer was named William Herschel and the new planet became known as Uranus.

William Herschel (1736–1822) was one of the most prolific astronomers and telescope makers of the 18th century. His accidental discovery of Uranus gained him international recognition and a royal stipend from King George III. One of Herschel's many projects was the cataloging of double stars. Following the consensus of the day, Herschel regarded close pairs in which one star was much brighter than the other as the chance superposition of two unrelated objects at greatly different distances. As Galileo first pointed out 140 years earlier, such pairs would be ideal candidates to search for stellar parallaxes. Herschel swept the skies with ever larger reflecting telescopes of his own design and manufacture in a search for such doubles, finally publishing a catalog of 269 pairs in 1782. After failing to detect any relative parallax shifts among these stars he put aside his double star project for several decades. It had been pointed out earlier by the Oxford professor John Mitchell that the type of stellar pairs found by Herschel were unlikely, on statistical grounds, to be unrelated chance occurrences but in fact were physical pairs.

In 1803 Herschel revisited some of his original double stars and discovered something unexpected. Several had noticeably altered their relative positions during the intervening 25 years. Herschel established that these changes could not be explained by proper motion. His search for stellar parallaxes had led him to the discovery of binary star motion. Such binary systems would for the first time allow previously unknowable properties of the universe and the stars to be determined. First, the fact that stars orbited one another established that Newton's laws of motion and law of gravity extended to the distant stars. Second, it clearly established that the stars were not at all similar to one another, since many binary systems contained stars of very different magnitudes and even colors. This realization effectively ended simple attempts to estimate stellar distances based on the assumption that they had the same brightness as the sun. Although not immediately realized, Herschel's discovery also held out the hope that from the study of the binary star orbits it would be possible to deduce the masses of the component stars from their orbital motions, in the same way it was possible to find the mass of Jupiter or Saturn from the orbits of their moons.

With photometric distance estimates at a dead end, the only remaining method of estimating the distances to the stars were direct methods based on stellar parallaxes. By the late 1830s astronomical instrumentation had been perfected to the point that attempts to measure annual parallaxes less than one second of arc were feasible for the first time. Moreover, several clues as to which stars might be the closest and show the largest parallaxes were available. Good candidate stars included very bright stars and stars with very large proper motions. A third indication of relative proximity included binary systems with well-separated components that also exhibited significant orbital motion. Sirius would easily qualify under the first two criteria but it was located too low in the southern sky to be precisely observed from most European observatories.

During the brief span of 18 months in 1837 and 1839, three observers working independently on three different stars and using different methods finally succeeded in measuring stellar parallaxes. Friedrich Georg Wilhelm Struve (1793–1864), working at Dorpat in modern-day Estonia made numerous attempts to directly detect the parallax shift of Vega between 1835 and 1836. In 1837 he reported a preliminary result but hesitated to claim a success, stressing that his parallax had a large uncertainty. He conducted further observations between 1837 and 1838 and published a final result of 0.2619 ± 0.0254 arc seconds in 1839 (the modern value is nearly 50% smaller).

Thomas Henderson (1798–1844), a Scottish lawyer and astronomer, made parallax measurements of several stars having large proper motions from the Cape of Good Hope beginning in 1832. Henderson left South Africa the next year but had amassed numerous measurements of the positions of both the star Alpha Centauri, the brightest star in the constellation of Centaurus, and Sirius. Both of these southern hemisphere stars were easily seen well above the horizon at the Cape. Henderson was in a position to attempt to measure the parallaxes with the data he had in hand, but he elected to wait till more observations were obtained with a better instrument by a colleague, who remained at the Cape. Back in Scotland Henderson finally completed his calculations of the parallax of Alpha Centauri and reported them at a January 1839 meeting to the Royal Astronomical Society. He obtained a result of 1.16 ± 0.11 arc seconds, about 35% larger than the modern value.

Twelve months later he published his parallax for Sirius. He concluded, "On the whole we may conclude that the parallax of Sirius is not greater than a half second of space; and that it is probably much less." In effect Henderson had reported a null result with an upper limit of 0.50 arc seconds. Nevertheless, a round value of 0.25 arc seconds, corresponding to Henderson's most likely estimate, was used by much of the astronomical community throughout most of the 19th century.

Wilhelm Friedrich Bessel (1784–1846) working in Königsberg, Prussia focused on the faint but rapidly moving star 61 Cygni. Called the "Flying Star" due to its rapid proper motion, 61 Cygni is actually a double star with a fainter companion

16 arc seconds away. Using an ingeniously devised telescope called a heliometer (Appendix A), Bessel worked from August 1837 to October 1838 carefully measuring and recording the positions of both stars with respect to one another and relative to two nearby field stars. He published his results in December 1838. In all, he obtained nearly one hundred measurements of the relative positions of both stars and subjected these observations to a rigorous statistical analysis. His final result was a parallax of 0.3136 ± 0.0202 arc seconds (the modern value is 0.2869 ± 0.0011 arc seconds). Interestingly Bessel also expressed the corresponding distance as 10.6 light years. In the end it was his relatively small 6% uncertainty and his exhaustive analysis that convinced most astronomers that the goal of directly measuring the distance to the stars had finally been reached.

Bessel was not content with simply measuring the distance to 61 Cygni, he used his result to also make the first estimate of stellar masses. Using his distance, Newton's law of gravity, and an estimate of the orbital period of 61 Cygni and its companion, he estimated the total mass of the two stars to be 2/3 that of the sun. His estimate of 540 years for the orbital period of 61 Cygni was in hindsight somewhat low. The modern value of the orbital period is 722 years, and this and the smaller parallax increases the total mass of the two stars to 1.1 solar masses. Nevertheless, Bessel had succeeded in directly measuring a stellar distance and stellar masses for the first time.

4

A Dark Star Prophesied

"But light is no real property of mass. The existence of numerous visible stars can prove nothing against the existence of numberless invisible ones"

—Wilhelm Friedrich Bessel, 1844

Newton's theory of gravity was a watershed in the development of science. It seemed to simultaneously provide a highly accurate and comprehensive means of both describing and explaining the motions of the heavenly bodies. Virtually all observed bodies moved according to Newton's basic rules. The new minor planets that were regularly being discovered after 1801 as well as those age-old harbingers of ominous events, the comets, were shown to obey the theory of gravity. Indeed the return of Halley's Comet in 1757, two decades after Edmund Halley's prediction, was widely touted as a monument to human intellect. Nothing in the heavens seemed beyond the scope of Newton's laws, even pairs of distant stars moved according to the same laws that governed the planets.

During the 18th and 19th centuries, elegant mathematical theories were constructed to account for a hierarchy of mutual gravitational effects between the planets and the sun. These elaborate mathematical models seemed capable of describing the smallest details of planetary motion. To keep pace with the predictive power of these theoretical constructs, better telescopes produced ever more accurate observations, which in turn increased the need for larger and more accurate star catalogs that provided the framework for charting these observations. There seemed no limit to the reach of Newton's laws. However, in the space of little less than two decades, between 1844 and 1862, three curious, and apparently unexplained aspects of gravitational motion emerged. These included the peculiar motions of the outermost and innermost known planets, Uranus and Mercury, and an anomalous but regular motion of the star Sirius. In their own way, each of these three paradoxical motions posed a similar challenge to Newton's gravitational edifice.

In essence, Newton's laws describe how a body will respond to the gravitational attraction of other bodies. When two (or more) bodies are evident then the corresponding gravitational motion is comprehensible. The calculations may be difficult or tedious, but in general nothing that is not observed, or at least known to exist, needs to be hypothesized: no extra assumptions are necessary. In each of the three gravitational anomalies discussed in this chapter, motions, without seeming explanation, were observed. In each instance, the resolution of the dilemma played a critical role in revolutionizing astronomy and physics. Moreover, to various degrees, all three dilemmas involved the most illustrious astronomer of the era, Urbain Jean Joseph Le Verrier. In what follows, these diverse problems, concerning the very outer and inner realms of the solar system and the distant stars, posed a similar theoretical impasse: something not seen was causing an observable effect. For reasons that will become obvious, these three seeming diverse dilemmas will be considered in the sequence of their ultimate intellectual impact.

Le Verrier (Figure 4.1) was born in 1811 in St. Lô, Normandy, where his father was a minor French provincial official. Although, Le Verrier was an excellent student, he was not admitted to the prestigious Ecole Polytechnique in Paris but had to settle on a lesser institution. He had intended to study chemistry but on the strength of his mathematical abilities, he was finally offered a position in astronomy at the Ecole Polytechnique in 1837, which he readily accepted. From the outset Le Verrier set himself some of the toughest problems in the solar system, including the search for a mathematical proof of the dynamical stability of the solar system itself. At the suggestion of François Arago, director of the Observatoire de Paris, Le Verrier took up the problem of Mercury's orbit for the first time in 1841. Although Le Verrier made great improvements in the determination of Mercury's motion, he was a perfectionist and withdrew his monograph while it was at the printer because his predictions of Mercury's transit of the sun on May 8, 1845 were off by a time interval of only 16 seconds. He abandoned Mercury, but would return to it again several times in his later career. This small, difficult to observe, innermost planet would provide Le Verrier with one of his greatest discoveries and also his greatest humiliations.

Le Verrier, however, was anything but a humble man. In spite of his modest upbringing he had a commanding and aristocratic bearing that seemed to fit naturally with his handsome good looks. He was precise and orderly both in his personal life and his calculations, and autocratic in his dealings with others. Le Verrier was also ambitious and a consummate self-promoter. Later, when Le Verrier had ascended to the pinnacle of French science, all acknowledged his talents, but there were few who personally liked the man.

Shortly after William Herschel's discovery of Uranus in 1783, it began to become apparent there were going to be problems establishing an accurate orbit for the new planet. The difficulties began when astronomers searched for, and found, a total of nineteen prior observations of Uranus, made during the previous 90 years, where the new planet had been mistakenly catalogued as a "star". For example, Bessel while reducing the positions in English astronomer James Bradley's stellar catalog noticed that Bradley had recorded Uranus as a "star" on December 3, 1753. These

Figure 4.1. Urbain Jean Joseph Le Verrier (photo by author).

earlier observations could not be easily reconciled with the post-discovery positions. Moreover, as Uranus continued to be observed, the situation only worsened. By 1820, not only were the older observations at considerable variance with theory, but evidence had accumulated that Uranus was exhibiting certain "irregularities" in its current motion, with the planet continuing to fall behind its expected path. This apparent failure of Uranus to precisely obey Newton's laws was a subject of great interest to mathematicians and astronomers of the period, for either it represented a flaw in Newton's theory or perhaps the influence of an even more distant planet. During the third decade of the 19th century, enough evidence had been amassed concerning the motion of Uranus that it became practical to answer the following question: If the perturbations of Uranus' orbit were due to the gravitational influence of a more distant unseen planet, would it be mathematically possible to deduce the location and mass of the perturbing planet? Of course, if astronomers had prior knowledge of the location and mass of such a planet, it would be straightforward to calculate its gravitational influence on Uranus. The inverse of this process, using the observed deviations of Uranus from its predicted orbit to locate an undiscovered planet, represented an entirely different problem of vastly greater difficulty, and a challenge to some of the most skilled mathematicians of the age.

As occurred with the determinations of the first stellar parallaxes, several individuals independently and nearly simultaneously arrived at a solution to the problem. In Königsberg, Bessel, as early as 1840, had reached the conclusion that the most probable explanation for the behavior of Uranus was the existence of some "unknown planet" and two years later he informed John Herschel he was preparing to take up the problem. Unfortunately, a fatal illness cut short his efforts. In England, John Couch Adams, a brilliant 24-year-old mathematician, then a fellow at St. John's College, Cambridge, also became convinced that the orbit of Uranus was being perturbed by an undiscovered planet and began working in earnest on the problem in the spring of 1843. Adams was encouraged in his efforts by the success of his initial calculations and by colleagues at Cambridge who also urged him to compete for the prize offered a year earlier, by the Royal Academy of Science in Göttingen, for a successful resolution of the problem of Uranus' motion. In France, two years later in June of 1845, Urbain J. J. Le Verrier, who was by now 34 years old and a recognized expert in the field of solar system dynamics, also took up the problem of Uranus' orbit at the urging of François Arago. At the time neither Adams nor Le Verrier were aware of the efforts of the other.

The calculations required deducing the mass and location of the hypothetical planet and were not so much conceptually difficult as they were laborious and unwieldy. It was first necessary to take the hundreds of observed dates and positions for Uranus recorded during the previous 60 years, carefully accounting for the influence of the other known planets, and then isolate those perturbations due to the putative planet. Then from these residual perturbations, it was necessary to first guess an approximate solution and proceed to refine that guess through arduous calculations, all the time being extremely careful, for something as simple as the accidental transposition of the digits in a key number could spoil weeks of effort. For nearly two years, such calculations filled the notebooks, first of Adams and then of Le Verrier. Adams finished first, but circumstances would conspire to deny him the honor of discovery.

In spite of Le Verrier's solid reputation, and the clear predictions he delivered to the French Academy of the location in the sky where the perturbing planet might be found, French observers were slow to take up the search. Impatient, Le Verrier, refined his predictions and sent his new coordinates to Johann Gottfried Galle, a young astronomer's assistant at the Berlin Observatory. Galle, and his immediate superior, the German astronomer d'Arrest, wasted no time in turning their telescope to the predicted place in the heavens and on the night of September 23, 1846, Galle was rewarded with the discovery of a lifetime, the eighth planet, Neptune, while Le Verrier laid claim to the eternal glory of having predicted it. Le Verrier was suddenly one of the most celebrated astronomers of the age. He also unsuccessfully lobbied behind the scenes to have the new planet officially named "Le Verrier".

The nearly simultaneous predictions of the new planet by Adams and Le Verrier sparked an intense cross-channel Anglo-French controversy for priority in theoretically predicting the new planet. The English felt that Adams had arrived at a solution first, while Le Verrier had simply been fortunate in sending his results to the right people. In contrast, the French regarded English assertions of Adams' predictions as an unjustified and belated attempt to diminish the achievement of Le Verrier. The dispute even spilled over into the popular press, where the national honors of both nations were fiercely defended. The British reaction was not entirely outwardly directed. Difficult questions were raised about the conduct of the Astronomer Royal, George Biddle Airy and his handling of Adams' predictions. Adams had made repeated attempts to enlist the support of Airy in a British search for the new planet. In hindsight, Airy could be viewed as being skeptical of the efforts of young Adams and being more impressed with results he was receiving from Le Verrier in Paris. It was argued that when Airy did finally react it was too little and too late. Ironically a British search, initiated by Airy in 1845, did actually succeed in recording the new planet but it was listed as a "fixed star" and the critical follow-up observation that would have revealed its movement came too late.

After the transcendent experience of theoretically predicting a major new solar system planet, Le Verrier quickly capitalized on his notoriety. In 1854 he was appointed director of the Observatoire de Paris, on the death of François Arago. There he busied himself with new more ambitious projects. Among these was the effort of tidying up the celestial mechanical details of the solar system. This task was neither incidental nor insignificant. Planetary dynamics had been painted in broad strokes and many details had been overlooked or glossed over. If Neptune was any indication, these details might well conceal other discoveries. The labor that Le Verrier set for himself was monumental. The motions of any particular planet were influenced, in principal, by the gravitational effects of all other planets. In practical terms this meant that the equations describing the motions of any one planet needed to include, not only the sun, but also the changing gravitational influences of all other bodies. Then these equations had to be compared with the hundreds of recorded observations and the unknown quantities determined. The latter phase of this effort involved long numerical calculations, which could be entrusted to careful but plodding individuals hired for the task. (Today this would involve writing rather simple computer programs.) Le Verrier's role was critical, he was the maestro. The calculations followed from detailed mathematical equations produced by Le Verrier, which needed to be written out, broken down into precisely defined calculations, and assigned to the hired staff. Over the years monographs on the motion of planet after planet emerged from Le Verrier's intellectual factory at the Observatoire de Paris.

In 1854 Le Verrier again returned to the planet Mercury, which had defeated him nine years earlier. Mercury being the planet closest to the sun moves the most rapidly, completing its orbit in only 88 days. It is also the most difficult to observe, spending most of its time in the glare of the sun. Additionally, Mercury has the most eccentric, or elongated, orbit of all the major planets. A circle has an eccentricity of zero and a completely flattened ellipse an eccentricity of one; Mercury has an eccentricity of 0.205. Highly eccentric orbits, such as Mercury's, mean extra work and more detailed computations since the equations which describe the orbit are usually expanded in powers of the eccentricity and extra terms are required to calculate quantities to a given precision.

As the problems with Mercury's orbit persisted Le Verrier determined that progress would depend on a laborious reworking of the orbital motions of the inner planets, Venus, Earth, and Mars. The earth's motion was critical since it is the body from which the position of Mercury is observed. As was the practice, the earth's motion was treated as the apparent motion of the sun. By 1858 Le Verrier squeezed the last vestiges of uncertainty from the motion of the sun and was ready to make a new attempt at Mercury.

As Le Verrier labored over Mercury he came upon a small but persistent discrepancy in its motion. The perihelion of Mercury's orbit was advancing too fast (see Figure 4.2). The perihelion is the point of the elliptical orbit closest to the sun and the

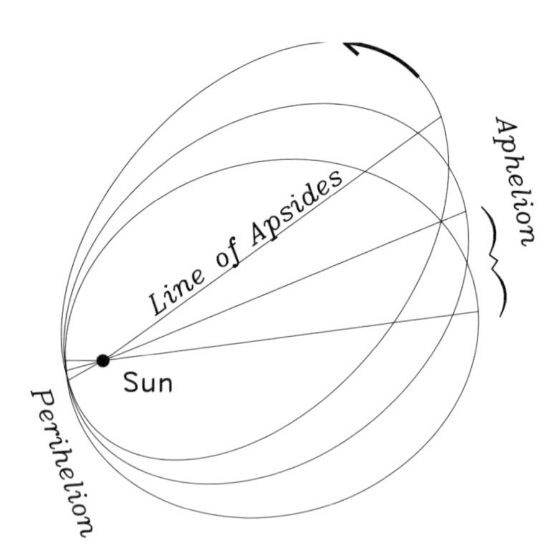

Figure 4.2. The perihelion advance of Mercury. The brace indicates the angular advance of the line of apsides. For clarity the eccentricity of Mercury's orbit has been exaggerated by a factor of four, and the rate of the advance of the line of apsides increased by a factor of 380,000.

aphelion, the most distant point. If a line is drawn from the perihelion, through the sun, to the aphelion, this line will bisect the orbit and it will remain fixed in space, if a planet, such as Mercury, moves only under the gravitational influence of the sun. However, Mercury is also subject to relatively small gravitational influences from all the other planets, principally from Venus, Earth, and Jupiter. These nonsolar gravitational forces perturb the purely elliptical orbit of Mercury and cause the perihelion to appear to advance through a small angle each orbit. This occurs because any gravitational forces not strictly due to a spherical sun will cause the orbit to fail to close upon itself after each cycle, resulting in a series of nearly identical, slightly overlapping and slightly shifted orbits. Le Verrier had the means to calculate these effects. However, every time he summed the contributions from the other planets he always came up short, Mercury's perihelion was advancing at a rate of 38 arc seconds per century too fast (the modern rate is 43 arc seconds per century) with respect to his calculated rate. This may not seem like much to worry about, since it would require about 3 million years for Mercury's orbit to come back to its original orientation, but it was well outside the limits of the uncertainty of either the calculations or the observations. Le Verrier announced his discovery of this inconsistency in September 1859.

Le Verrier's experience with the perturbations of Uranus' orbit provided him with a natural explanation; there could be an undiscovered planet orbiting between the sun and Mercury, causing the unexpected behavior in Mercury's orbit. He calculated what sort of planet would be required and determined a range of possible masses and orbital radii. Could such a planet have escaped detection and if so, could it be found if an effort were made to look for it?

On average, Mercury is observed to cross the face of the sun about thirteen times a century: the last such event occurred on November 8, 2006. These transits provide the best fixes on the planet's position in its orbit and were used by Le Verrier to constrain his calculations. During these events, Mercury appears as a small black dot crossing the solar disk in a few hours. Any planet interior to Mercury might also reveal itself with such a transit, if it could be distinguished from common sunspots. Another way the elusive planet might be found is during a total solar eclipse, when it might be seen shining brightly near the sun, if it could be distinguished from known background stars.

At about 4 p.m. on the afternoon of March 26, 1859 a provincial doctor and amateur astronomer named Lescarbault had witnessed a small black dot near the sun's visible limb, with an apparent motion that could not be easily attributed to a sunspot. Le Verrier first became aware of Lescarbault's observation when the doctor wrote to him in late December of 1859. When Lescarbault's letter reached him on December 30, Le Verrier immediately took the afternoon train from Paris and questioned the village doctor in person about what he had seen through his small telescope. Satisfied that the man was credible, Le Verrier announced the observation, noting that it could be the unseen planet he had hypothesized three months earlier. Although Le Verrier was cautious in his pronouncements, considerable public excitement was generated and the "planet" acquired both a name "Vulcan", and more reality than was justified. Just a single observation did little to diminish the calculated range of possible orbits for Vulcan. Old sunspot records were searched for other potential sightings and several candidate events emerged. Using these, Le Verrier made new calculations and issued predicted times of future transits of Vulcan and likely positions during forthcoming solar eclipses. Much of the astronomical community enthusiastically joined in the hunt, but when the predicted transits failed to occur and total solar eclipses yielded only occasional disputed sightings, interest declined. Over the next twenty years additional unexpected transits of Vulcan were reported and these prompted yet more predictions from Le Verrier and others, but nothing solid ever materialized, and the reality of Vulcan lost favor with most astronomers.

If Vulcan was not real, the advance of Mercury's perihelion certainly was, and other possible explanations were put forward. Perhaps there was not one large Vulcan but many small ones or even a ring of material interior to Mercury's orbit. However, there was never any observational evidence for such bodies. It was even proposed that Newton's inverse square law of gravitation required a small *ad hoc* modification to explain Mercury's motion. When Le Verrier died in 1876 the mystery still remained. The dilemma was finally resolved in 1916 when Albert Einstein showed that his newly developed theory of General Relativity (Chapter 9) could naturally explain Le Verrier's discovery. As we will see, in 1862, at the peak of Le Verrier's involvement with Vulcan, he was also actively pursuing the era's third gravitational puzzle, the motion of Sirius.

On the 10th of August 1844 Friedrich Wilhelm Bessel in Königsberg (Figure 4.3) sent a letter to Sir John Herschel in England. Herschel promptly translated the letter and published it in the *Monthly Notices of the Royal Astronomical Society*. In the letter Bessel reported a rather disturbing discovery regarding the proper motions of the two stars, Sirius and Procyon. He had been systematically determining the positions and proper motions of a series of "fundamental" stars in an attempt to build up a fixed frame of reference for the celestial coordinates of all stars. To do this he had analyzed and compared the apparent positions of a number of stars from different observers in Greenwich, Palermo, Pulkovo, Königsberg, and Cape Town over a 90-year period. The assumptions on which he based this effort were simple enough, the stars had constant, *rectilinear*, proper motions, and if one carefully and consistently accounted for the precession of the equinoxes, stellar aberration, refraction by the earth's atmosphere, and a host of instrumental and observational effects, peculiar to each observatory and observer, then it would be possible to define a "fixed" reference frame for the entire sky using only a small number of "fundamental" stars. With respect to this frame of fundamental stars, the positions and motions of all stars could be estimated and predicted for future dates. In effect, this ideal was more or less guaranteed by Newton's laws of motion. All that was necessary was to determine the position of a star in this frame at two well-separated dates and its proper motion would be determined. The key assumption was that the stars, being distant and having small proper motions, would appear to travel paths in the sky that are well approximated by straight lines.

The observations that particularly concerned Bessel were those of his "fundamental stars" Sirius and Procyon. A decade earlier he had first noticed that Sirius in particular did not appear to fit this simple conceptual model and he opened his letter

Figure 4.3. Wilhelm Friedrich Bessel (Marcus Kaar).

with a direct statement of the problem: "The subject which I wish to communicate to you, seems to me so important for the whole of practical astronomy, that I think it worthy of having your attention directed to it. I find namely, that existing observations entitle us without hesitation to affirm that the proper motions, of Procyon in declination and Sirius in right ascension, are not constant; but, on the contrary that they have, since the year 1755, been very sensibly altered." Bessel then went on to describe in detail the problems that he found with the proper motion of these two stars. In demonstrating the problem, he avoided the use of the absolute positions of the individual stars, instead focusing on their differential motion with respect to the mean motions of several other "fundamental stars". In the case of Sirius, he listed the differentials in right ascension, the east–west angle, with respect to the stars α and β Orionis and α Canis Minoris (Procyon) in tabular form for a series of dates between 1755 and 1843. These differences appeared to change sign over time and to be too large to attribute to instrumental or other effects. Moreover, since these were differential measurements, his results were relatively immune to systematic effects associated with any single observatory. Bessel's discovery was particularly disturbing because of the

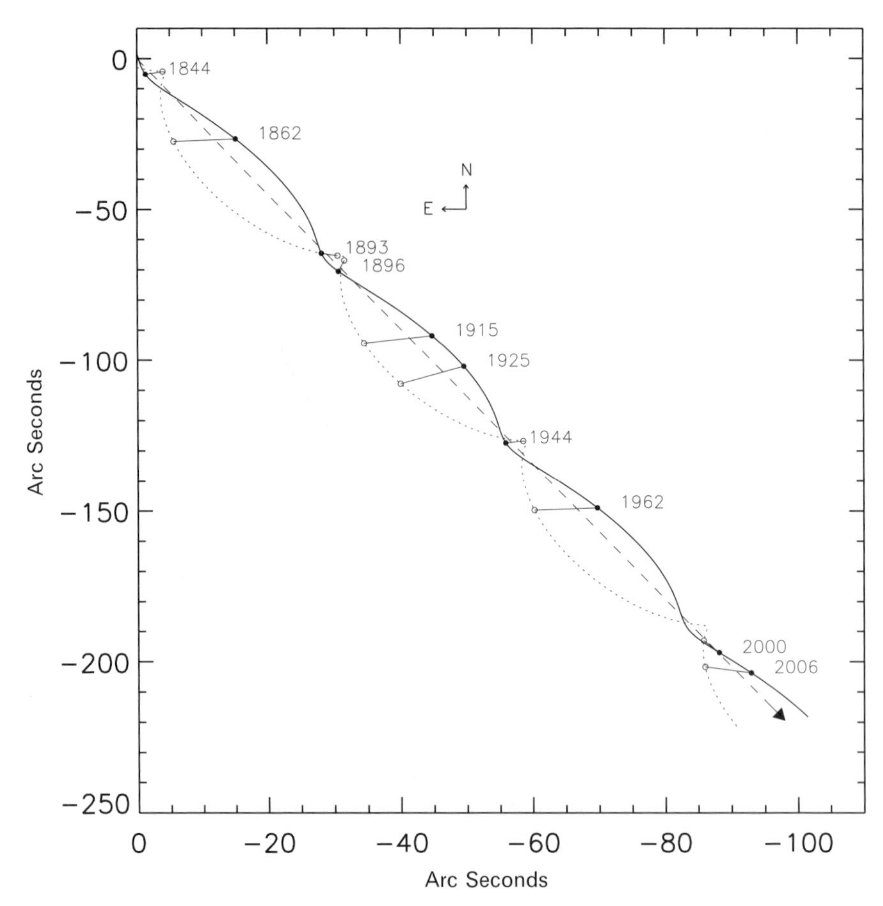

Figure 4.4. The anomalous proper motion of Sirius. The dashed straight line represents the path of the Sirius system, while the solid sinuous line is the path of Sirius. The dotted sinuous line is the path of the unseen companion. The solid and open circles indicate respective relative positions of Sirius and its companion during key dates.

challenge it posed to the goal of employing constant proper motions to establish an absolute celestial frame of reference defined using a limited number of observations of each star. He states, "If this is so, the observations of the place of a star at two epochs are no longer sufficient to express its place at any definite time ... It follows also from this, that we are yet very far off from the correctness we have imagined ourselves to have arrived at in the fundamental determinations of astronomy; and, that a new problem presents itself, whose solution will cost much labor and a long period of time." Obviously if such prominent stars as Sirius and Procyon show perturbed motion, other stars may also and the observational effort needed to account for such effects would require a much more extensive observing program extending over many more years (see Figure 4.4). This was not particularly good news for global astrometry.

The part of his letter that has had the most lasting impact was not the potential problem this posed for measuring the positions and motions of stars, but the conclusion he reached regarding the nature of the perturbation. He assumed gravitation was the force perturbing the motions of Sirius and Procyon and went on to develop a simple expression for the observed deviations in terms of the relative mass and location of the perturbing body or bodies. Bessel examined and rejected several hypotheses such as a disturbing mass near the sun or the "joint action of millions of existing stars" as being unable to account for the observations. In the end he focused on unseen masses, which must be very near Sirius and Procyon and concludes that these stars are part of "smaller systems". "If we were to regard Sirius and Procyon as double stars, the changes in their motions would not surprise us; we should acknowledge them as necessary and have only to investigate their amount by observation." But if this was the correct explanation, why were the hypothetical companions not seen, as they were in other binary systems? Bessel's answer would in later years prove very perceptive: "But light is no real property of mass. The existence of numberless visible stars can prove nothing against the existence of numberless invisible ones." He concluded with the observation that these perturbed stellar motions may "provide important information concerning the physical constitution of the universe." The German/Estonian astronomer Johan Mädler in 1867 declared Bessel's discovery to be "the foundation of an Astronomy of the Invisible."

Bessel's deduction of unseen objects in orbit about the two "dog stars", Sirius and Procyon, would be a powerful demonstration of the deductive power of Newton's theory of gravity, if the "companions" could be found. One method that was available to astronomers was to continue to study the motion of Sirius and Procyon in the hope of learning more about the nature and relative location of the unseen companions. As in the case of Neptune, perhaps the observed perturbations could provide indications of what to look for and where to search for it in relation to the primary star.

After several months of illness Bessel succumbed to cancer at Königsberg in March of 1846, never witnessing the results that would flow from his original insight concerning Sirius and Procyon. He was followed as director of Königsberg by Christian August Friedrich Peters (1806–1880). In his 1851 habilitation thesis, Peters systematically extended Bessel's original analysis of the motions of both Sirius and Procyon. He included additional observations up to 1848 and undertook a more thorough analysis of the corrections that needed to be applied to the observations made by various observatories. Peters succeeded and was able to be much more specific about the nature of the companions of both stars that Bessel had discovered. For Sirius, Peters found that the companion had a period of 50.093 years, and moved in a highly eccentric orbit, that the unseen body had most recently passed near Sirius in 1841. More importantly he produced an ephemeris, or table of predicted positions, for Sirius listing the deviations of Sirius' motion for the years 1755 through 1870. The latter information was particularly important since it gave astronomers a first hint of where, with respect to Sirius, the unseen body might be found and some criteria for distinguishing any star-like object that might be discovered, from unrelated background stars that Sirius chanced to pass near as it moved through the heavens. However, only very general indications about location and mass of the companion were possible, since only the one-dimensional motion of Sirius was observed and not that of the companion. Nevertheless, what Peters had achieved was a large improvement over the mere knowledge that something was orbiting the brightest star in the sky. Peters was even able to place a crude limitation on the mass of the unseen body; it had to be *greater than* six times the mass of Jupiter. How much greater could not be determined from existing data, but it could not be less.

Although Peters' results provided the clearest existing indication of the perturbed motion of Sirius, the picture remained one-dimensional since again only east–west right ascension motion was considered. To obtain a more complete picture of what was occurring, it would also be necessary to determine how Sirius was moving in the north–south, or declination direction. The observations existed but such an investigation was made difficult because most data were from European observatories, where Sirius never rises very far above the horizon. Under such circumstances starlight undergoes a strong refraction (Appendix A), or bending, due to the earth's atmosphere. This bending causes the position of a star to appear farther north, at a higher elevation, than its true position. For example, when Sirius is on the meridian at Königsberg, depending on the temperature of the air, its apparent position is some 167 arc seconds north of its true position. Although this effect can be calculated and corrected for, it is quite large compared with the motion of Sirius and comparing results from different observatories at different latitudes is a difficult task requiring subtle analysis.

By 1861, however, several attempts were being made to include the north–south motion of Sirius. Such a determination would provide two new important clues as to the mysterious companion. The first clue would be the location of the center of mass between Sirius and its unseen companion. The center of mass is an empty point in space about which Sirius and its companion appear to orbit. It is the center of mass whose proper motion moves in a straight line with respect to the background stars. The center of mass for the stars in a binary system can be thought of as the fulcrum or balance point along the line joining the centers of the two stars. The second clue, of more immediate interest, was the direction, with respect to Sirius, where the unseen companion might be found. If an imaginary line is drawn from Sirius to the center of mass, then the dark companion must lie along the projection of that line beyond the center of mass. Its separation distance from Sirius would not be known but at least astronomers would possess a clear indication of the direction in which the predicted companion might be expected to lie.

In a situation that in many ways mirrored the independent efforts, of Adams in Cambridge and Le Verrier in Paris, to theoretically predict the location of the planet Neptune in 1846, a theoretical search for Sirius' companion was independently undertaken in 1861 by two talented young astronomers. One effort was due to Truman Henry Safford (1836–1901), then an assistant at the Harvard College Observatory in Cambridge, Massachusetts. As a child Safford (Figure 4.5) was well known throughout New England as a mathematical prodigy with a remarkable gift for computation, which earned him the title the "lightening calculator". Safford was one of those rare individuals who could mentally multiply and divide large numbers and extract square and cube roots with astonishing ease. A clergyman who inter-

Figure 4.5. Truman Henry Safford, circa 1844 (permission Harvard University Archives).

viewed him at age ten described how, when a difficult problem was posed, the boy would shoot out of his chair "like an arrow" and fly around the room, "his eyes wildly rolling in their sockets" before stating the answer. At the age of eleven Safford was publishing his own almanac and by fourteen he was said to have computed the orbit of a comet. In 1847 he visited the newly opened "Great Refractor" at Harvard College to view the new planet Neptune that had been discovered only the year before. Eleven-year-old Henry, very much satisfied with his view of the new planet, termed the experience "great sport". Safford graduated from Harvard College at eighteen and joined the staff of the observatory as an assistant to the director William Cranch Bond and later as assistant to Bond's son, George Phillips Bond.

In August 1861, at the age of 25, Safford had completed his calculations showing that the north–south declination motion of Sirius was completely consistent

Figure 4.6. Arthur Auwers (Mary Lea Shane Archives of the Lick Observatory, University of California-Santa Cruz).

with east–west right ascension motion determined by Bessel and Peters. He sent a lengthy description of these results to the University of Michigan publication *Brünnow's Astronomical Notices*, where it appeared on December 20th, 1861. He also prepared an abbreviated report for the *Monthly Notices of the Royal Astronomical Society* that would appear in the spring of 1862. As Safford was doing this, a second determination of the north–south motion of Sirius, was being independently completed by 24-year-old Arthur Auwers (1838–1915, Figure 4.6), a doctoral student in Königsberg, following in the footsteps of Bessel and Peters. Auwers, who was well along in his own calculations, became aware of Safford's report in *Brünnow's Astronomical Notices*. This prompted Auwers to submit his own paper to the *Monthly Notices of the Royal Astronomical Society*. Both men simultaneously published summaries of their results, side by side, in the March 1862 issue of the journal. The publication of these two efforts, which now allowed a full two-dimensional determination of the orbit of Sirius about its center of gravity, was to prove to be uncannily timely, as events in Cambridgeport, Massachusetts and Paris would soon show.

In Paris, Le Verrier, whose domain of expertise was the gravitational motion of bodies in the solar system, was very much aware of the stellar dilemma posed by Bessel's unseen companion of Sirius. As early as 1855, he had suggested that the companion might be detected with an improved telescope and implied that it could turn out to be a large faint planet orbiting Sirius. In 1862 as the director of one of the world's premier observatories, Le Verrier commanded the resources and instruments to mount a search for this mysterious body. Fortuitously for Le Verrier, the Observatoire de Paris had just come into possession of a unique instrument that might reveal Sirius' hidden companion.

Among the staff at the Observatoire de Paris was the ingenious Léon Foucault (1819–1868), who is chiefly remembered for the Foucault Pendulum, which demonstrates the rotation of the earth, the gyroscope, and for the first accurate terrestrial measurements of the velocity of light and thereby the distance to the sun. Foucault, the observatory physicist, had been asked by Le Verrier to build what would have been at the time the world's largest refracting telescope with a diameter of 74 cm (29 inches). Foucault convinced Le Verrier that a reflecting telescope with a glass primary mirror, rather than a traditional metal mirror, would be the easiest option. After building several smaller prototypes Foucault completed an 80 cm (31.5 inches) reflecting telescope using a new technique of chemically applying a thin layer of metallic silver directly onto the polished front surface of the mirror. To figure his larger mirrors Foucault also developed a simple optical test to reveal imperfections in the mirror surface, so that they could be removed. Foucault's revolutionary glass reflector was much superior to the metal reflectors then in use and a match for any of refractors of the era; it also set the stage for today's large reflecting telescopes.

The new telescope went into service in mid-January 1862. One of the first to peer through the new instrument was Le Verrier himself, the serene theorist, who discovered new planets "with the tip of his pen, without other instruments other than the strength of his calculations alone, . . ." Some time, on or just before Monday, January 27, 1862, deep in the glare surrounding Sirius, Le Verrier thought he saw something.

5

A Dark Star Revealed

Thus we have added to the catalog not a new star only, but a magnificent wonder; and we may still be reminded of one of the last remarks of La Place, and certainly a very impressive one: "That which we know is little, that which we know not is immense"

—George Philips Bond, 1863

In early 1843 a spectacular comet appeared in the sky. It grazed the surface of the sun on the last day of February, and in early March, as it raced back towards the outer regions of the solar system, it put on a brilliant display. For those in the northern hemisphere who witnessed the comet, it presented a glorious tail that stretched in an arc of 20 to 45 degrees across the southwestern sky after sunset. By all reports it was one of the brightest comets of the 19th century, clearly visible near the sun during daytime. It came to be known as the "Great March Comet" and was seen by large numbers of the public in North America where it provoked much comment and discussion. One tangible result of the Great March Comet was that some of the more prosperous citizens of Boston, Massachusetts were moved to contribute, by way of public subscription, the considerable sum of $25,000, towards the construction of a large refracting telescope, equal to any in Europe. The most critical component of the telescope was the 15-inch diameter objective lens contracted from the German firm of Merz & Mahler in Munich at a price of $12,000: there being no source of a lens of such size and quality in the United States of that day. The lens was the twin of one produced earlier for Friedrich Wilhelm Struve at the Pulkovo Observatory in Russia. The new observatory was completed in 1847 and located on Observatory Hill on the Harvard College campus. The director of the new observatory was William Cranch Bond (1789–1859), a Boston clock maker, who had been appointed in 1839, at no salary, by the Harvard Corporation as "Astronomical Observer to the University". One of Bond's first achievements with the new telescope was the discovery of a new moon of Saturn, Hyperion, in 1848. The "Great Refractor" of Harvard in Cambridge, Massachusetts remained one of the two largest refracting telescopes in the world for the next twenty years.

Shortly after the "Great Refractor" was installed, Director Bond received a request from a Boston portrait painter and tinkerer named Alvan Clark who desired to see the telescope and to be able to look through it. Bond was familiar with Clark, for on several occasions he had examined reflecting telescopes produced by Clark and his sons, finding them "quite inferior". Bond politely informed Mr. Clark that he would need to obtain a written letter from the university president in order to honor such a request. Clark, who had painted portraits of such prominent Bostonians as Daniel Webster, seems to have had no trouble in obtaining the required letter and arrangements were made for him to inspect the telescope. Mr. Clark's interest, it turns out, was much more than idle curiosity and his frank assessment of the telescope and the quality of the optics must have astonished, or perhaps annoyed, Bond. As Clark later wrote in his brief autobiographical sketch, "I was far enough advanced in knowledge of such matters to perceive and locate the errors of figure in their 15-inch glass at first sight, yet those errors were very small, just enough to leave me in full possession of all the hope and courage needed to give me a start, especially when informed that this object glass alone cost twelve thousand dollars."

Alvan Clark never set out to build telescopes, if anything he aspired to be a millwright like his father. He was born in 1804 in Ashford, Massachusetts, the fifth son of ten children, into the family of a New England farmer and mill owner. He had only a primary school education and labored on the farm and the family's grist and saw mills until age 17, at which time he briefly worked for a wagon maker for a year. He returned to the family farm and "... put myself at work in good earnest to learn alone engraving and drawing." In 1825 he happened to ask a man traveling to Boston to return with some sable hair brushes. When the brushes arrived, Alvan noticed an advertisement in the newspaper wrapping headed: "Engravers Wanted". He quickly traveled to Boston but found that the firm, which printed calico for the New England mills, had already contracted for the engraving with a Philadelphia company. Fortunately, one of the Philadelphia firm's representatives was visiting and offered Clark a job, on the spot, at eight dollars a week with the opportunity to learn the engraving trade. Clark was quickly promoted and moved with a partner of the engraving firm to Providence, Rhode Island, but there it seems that the master engravers, who were imported from England, jealously guarded their techniques. Alvan was offered and accepted a position with the firm's branch in New York City. It was in New York that Alvan Clark seems to have blossomed as an artist. In the time he could spare from his engraving, he studied art and by 1829 was exhibiting his work at the National Academy of Design, which had been founded by Samuel F. B. Morse, of telegraph fame. His specialty was miniature portraits, popular in the era before photography. By 1839 he had given up the engraving trade, concentrating on his portraits and making a comfortable living charging $20 "a head".

Throughout his entire life Alvan Clark possessed keen eyesight, and a strong attention to detail coupled with a practical urge to improve upon all manner of things that caught his interest. Quite apart from his artistic talents, Alvan Clark was also well known in New England as a skilled marksman with a musket. It was said he could repeatedly place rounds in the hole made by his first shot. He maintained that he could teach any novice to be a crack shot after only a few minutes of instruction and five shots: "even ladies" can hit the target with "all the perfection imaginable" he wrote. He built his own rifles and methodically experimented with them in an effort to improve accuracy. As a consequence, he recognized that one of the principal defects of the muzzle-loading rifles of the day was wear and damage to the muzzle and the conical lead projectile arising from the action of loading with the ramrod. His ingenious, but practical, solution was to cut off an inch or so of the muzzle and then to reattach the segment with four dowel pins, creating a "false muzzle". In this way, loading could be done with the "false muzzle" in place and then the gun fired with it removed. This preserved the end of the muzzle and eased the loading process, resulting in a significant improvement in accuracy and repeatability, provided one remembered to remove the "false muzzle" before firing! In 1840 Clark patented his design, which was one of the last major improvements in muzzle-loading rifles before the advent of breach-loading firearms. He later sold his patent rights to the company of Edward Wesson; receiving the sum of $2 for each rifle made with his design.

Alvan Clark's career as a world famous optician and telescope maker began quite by accident when the dinner bell at Phillips Academy in Andover, Massachusetts broke in 1844. At the time, Alvan's oldest son George Basset Clark (1827–1891), a student at Phillips studying to be a civil engineer, had become interested in building a reflecting telescope. He collected the metal shards from the bell and brought them home to Boston where he melted them down, together with some tin, on the stove in the family kitchen and cast a five-inch disk of the metal. The elder Clark keenly watched his son's progress in grinding and polishing the curved mirror surface. Alvan recalls that "I was at some pains to acquaint myself with what had been done, and how done, in this curious art," in an effort to assist his son. A successful telescope was soon completed and the father and son used it to view the moons of Jupiter and the rings of Saturn, among other sights. Intrigued by the possibilities, other larger and better telescopes soon followed. Clark Sr. used some of these to sketch maps of objects he observed. The quality of these early instruments can perhaps be gauged by the fact that some of the stars recorded by Clark in Orion had been overlooked by Sir William Herschel using a much larger telescope. Experiments with reflecting telescopes continued for several years until Clark convinced himself that the difficulties involved with the metal mirrors of the time were "irremediable". At some point he proposed to his son that they try building a refracting telescope. George, however, doubted they would succeed, since textbooks described the grinding of lenses as very difficult.

The father and son must have met with success, as a series of larger and larger refracting telescopes were soon being sold to interested parties in the Boston area. A small firm, Alvan Clark & Sons (Figure 5.1), consisting of Alvan Sr. and his sons George Bassett and Alvan Graham, was founded in 1851. George took over the mechanical aspects of the instruments, such as mountings and clock drives, while Alvan Sr. and Alvan Jr. focused on the figuring and polishing of the lenses. The recognition that Alvan Clark & Sons was capable of turning out first-class instruments came in that same year. Alvan Sr. had written to the Revd. W. R. Dawes in

Figure 5.1. Alvan Clark and sons. Left to right, George Bassett Clark, Alvan Clark, and Alvan Graham Clark (Mary Lea Shane Archives of the Lick Observatory, University of California-Santa Cruz).

Haddenham, England, reporting on his activities and mentioning two new double stars he had observed. Although a clergyman, Dawes was also a respected double star observer who had a reputation as a keen judge of telescopes and their optics. Dawes replied, and as Clark later said, "... my correspondence with Dawes had become more extensive than with any fellow mortal in all my life." In 1853 Clark had finished a $7\frac{1}{2}$-inch objective lens and used it to discover a close companion to the star 95 Ceti. Learning of this, Dawes inquired about the possibility of acquiring the lens. Before a deal was struck, however, Dawes sent Clark a list "... of Struve's difficult double stars, wishing me to examine them, which I did and furnished him with such a description of them as satisfied him that they were well seen." The lens was sold and four others followed, including an 8-inch. Dawes was obviously pleased with what he was receiving from his unlikely source in Boston, and he soon became a valuable exponent of Clark, touting the quality of his lenses in the pages of the *Monthly Notices of the Royal Astronomical Society*.

In 1859 Alvan Clark traveled to England to visit Dawes and to deliver two 8-inch objective lenses. Characteristically, Clark notes with some irritation, that the finished lenses passed though Liverpool customs duty free, while he had been forced to pay a duty when he imported them as unfinished glass blanks into the U.S. This remark refers to a long-standing dispute that Clark had with the Boston customs inspector, a "Mr. Austin", who insisted Clark pay 30% duty on a rough glass blank that turned out to be flawed and worthless. Clark was then forced to pay another 30% duty to import a replacement. As Clark, a confirmed Yankee and supporter of the Union cause, concluded in his autobiography, "But then we were under a Democratic administration." Years later Clark fondly recalled his transatlantic visit, including a trip with Dawes to Greenwich Observatory in London and to a meeting of the Royal Astronomical Society where Clark was able to meet and talk with "many notable personages" such as Sir John Herschel and Lord Rosse.

Clark's greatest challenge came in 1860, when Dr. Frederick A. P. Barnard, president of the University of Mississippi in Oxford, Mississippi, commissioned Alvan Clark & Sons to build the then largest refractor in the world, with an $18\frac{1}{2}$-inch objective lens. This new commission required a larger workshop and so the family home was sold and money borrowed to purchase an acre and a half of land near the Charles River in Cambridgeport. The glass blanks were ordered and work was begun in 1861, with the Clarks promising to have it ready for Dr. Barnard to inspect and look through by June of 1862. By this time, however, the Civil War had led to a complete severing of commercial relations between the northern and southern states and the project was all but forgotten in a Mississippi bent on secession from the Union. The lens was never paid for. The Clarks, having invested much of the firm's capital and reputation in the massive lens, were determined to complete the figuring process. It was during the testing of this lens in early 1862 that they were to make their most famous discovery.

The early evening of Friday, January 31, 1862, looked promising as George Phillips Bond (1826–1865) "disrobed" Harvard's Great Refractor for an evening of observations on Observatory Hill. G. P. Bond was the son of William Cranch Bond, having succeeded his father as director of the Harvard Observatory in 1857. Observing conditions in Cambridge over the prior two weeks had been barely tolerable. As he pointed the telescope east towards his favorite target, the Orion Nebula and its embedded stars, Bond anticipated that it would be a night of good observing. It was around 8 p.m. and Orion was climbing higher in the sky, as Bond was busy studying some nebulosity near the stars θ^1 and θ^2 Orionis. At the same time, across Boston, at their home and workshop on the banks of the Charles River, Alvan Clark and his son Alvan Jr. were using the opportunity of this cold but clear night to test the $18\frac{1}{2}$-inch lens or "object glass", as some called it. It was common practice for the Clarks to field-test the resolving power of their lenses on double stars and to use bright blue stars to test the color correction. During such tests the lens cell was mounted at the end of a crane-like boom in the yard of their workshop. On this particular night the senior Clark was also pointing his instrument to the eastern sky, somewhat south of Orion. He had selected Sirius to color-test the lens, but was having trouble steadying the telescope. His son took over just as Sirius was clearing the roof of a nearby building and after a few seconds he noticed a very faint star in the glare of Sirius. "Father," he said, "Sirius has a companion." The elder Clark quickly confirmed what Alvan Graham had seen. It is doubtful if either the father or son were aware at the time of the significance of the faint companion they sighted that night, or in particular of Bessel's earlier predictions of its existence. It is likely they were not, since they left no written record of the historic discovery. Nevertheless, the new star was apparently shown to Dr. Henry Coit Perkins, a prominent New England physician and early Daguerre-type photographer. Alvan Clark, however, had previously discovered a number of double stars with his telescopes and was in the habit of routinely reporting such discoveries both to the local newspapers and to professional astronomers.

Word of the discovery began its journey around the world the next day. The first stop was just across Cambridge, at the Harvard College Observatory. The observatory logbook for the next night, Saturday, February 1, records no observations but does contain the following entry in George Bond's handwriting; "A. Clark jr. reports the discovery of a companion of Sirius last evening with the $18\frac{1}{2}$ in. object glass. Following in AR at 10″ dist." The latter comment refers to the approximate orientation and separation of the two stars. It is likely that Alvan Graham had written a brief report of the discovery on a piece of writing paper, folded into a square packet, bound it with a string, and had it delivered to Bond the following Saturday morning. It would require a full week before the skies over Cambridge permitted Bond to confirm the discovery. However, on February 7, Bond notes "After scrutinizing the neighborhood of Sirius on every clear night of the week since Feb. 1 without obtaining a single opportunity of quiet definition a sight of [the] companion was attained this evening during an interval of about 5^{m} when the vision was tolerably good. For an hour before and after it was impossible to catch a glimpse of it." Next to the entry is a tiny sketch representing what he observed through the eyepiece. It consists of an arrow pointing to the right, indicating east, together with a large dot and a much smaller dot to the left. Bond, conscious of the instability of the "seeing" that evening, immediately set to work measuring the angular separation and the clock angle between Sirius and the faint companion. He measured the anti-clock angle as 88°55′, east of north, and visually estimated the angular distance at 12 arc seconds. A short time later, when he recorded the vision as deteriorating rapidly, he quickly made two measurements of the separation with the micrometer, obtaining a mean of 11.36 arc seconds. His final comment was, "A very difficult object when the vision is at all distorted." Although Clark's companion star had now been confirmed and measured for the first time, it would require some time, a great deal of observation, much computation, and a certain amount of confusion to finally determine if Clark's discovery was in fact Bessel's dark companion. Bond, however, was confident that the mystery had finally been solved.

George Bond had further opportunities to measure the position of the faint companion star on several more occasions that spring before Sirius became lost to the evening skies of April. His next observations were on February 10th and 16th. On February 18th Bond was invited by Alvan Clark Sr. to the family workshop in Cambridgeport to view Sirius and its companion through the $18\frac{1}{2}$-inch lens. Bond recorded the events of the evening in the logbook of the Great Refractor under the heading "Sirius Clark $18\frac{1}{2}$ Object glass".

"Saw Sirius with its companion at about 8 p.m. perfectly distinct was much improved with the excellent definition of Sirius with the $18\frac{1}{2}$ in object glass. It is difficult to suggest in any particular which improvement can be desired." He went on to give his professional assessment of the color correction properties of the new lens.

"When the eye piece is drawn out beyond focus the outer fringe of the image is faint green. For bluish stars the color correction seems very perfect. In general there is a slight excess of red but not at all offensive." Later in the evening they examined the star Rigel and then Bond had an opportunity to view his beloved nebulosities in Orion, as he had never observed them before. Bond was impressed enough to include a glowing description of what he had seen in Orion with the Clarks' object glass in the annual *Harvard College Observatory Report* for the following year of 1863.

Only two days after his observations of February 10, Bond mailed a report of the discovery to the *American Journal of Science*, a bimonthly journal, also known as *Silliman's Journal*, which published reports on a variety of natural history topics submitted by scientists and enthusiastic amateurs in early 19th century America. In this report Bond states that "It remains to be seen whether this will prove to be the hitherto invisible body disturbing the motions of Sirius, the existence of which has long been surmised from the investigations of Bessel and Peters upon the irregularities of its proper motion in right ascension." He then takes the opportunity to favorably mention the very recent work of his young assistant Truman Henry Safford, whose calculations of the north–south declination motion also confirmed the results of Bessel and Peters, but cautions that it will require several years of observation to ensure that the newly discovered star is a true physical companion of Sirius and not just a random background star. He notes, however, "For the present we know only that the direction of the companion from the primary accords perfectly with theory." By this he means the work of Bessel, Peters, and now Safford. He concludes with the following statement, "Its faintness would lead us to attribute to it a much smaller mass than would suffice to account for the motions of Sirius, unless we suppose it to be an opaque body or only feebly self-luminous." He was expressing his belief that the new star is either shining by reflected light, like a planet, or it is a star of remarkably low luminosity.

News of the discovery quickly reached New York City, where Mr. Lewis Morris Rutherfurd read of Bond's description of the discovery in the March issue of the *American Journal of Science*. Rutherfurd was prominent in the early application of spectroscopy and photography to astronomy. He promptly succeeded in observing the companion on March 8th with an $11\frac{1}{2}$-inch refractor at his private observatory located in the back garden of his house on the corner of 2nd Avenue and 11th Street, in the present-day Bowery area of Manhattan. He immediately wrote to Bond inquiring as to where the orbital calculations of Safford were published, since he had noted a change in the position with respect to that reported by Bond, and added a note saying he was sending Bond a copy of the stellar spectra he was obtaining with his recently fabricated spectrograph.

Simultaneously with his report to the *American Journal of Science*, Bond also sent a very brief account of the February 10 observations to the *Astronomische Nachrichten*, the premier astronomical journal of the time in continental Europe. The report crossed the Atlantic by packet steamer arriving several weeks later on the desk of the editor, none other than Dr. C. A. F. Peters (Chapter 4). Peters, who a decade earlier had greatly improved on Bessel's study of the motion of Sirius, was apparently not immediately convinced that Clark's discovery corresponded to the expected position of Bessel's dark star, nevertheless he promptly placed Bond's report in the March 21 issue, without comment. Among his peers in Europe, Bond was much less speculative. He merely noted that he had confirmed the Clarks' "interesting discovery" and provided its location relative to Sirius. There was no mention of Safford's work or that this might finally be the unseen companion of Sirius. Later that summer, Bond would report eleven additional observations of Sirius and its companion that he conducted during the months of January to March of 1862. It is in this second *Astronomische Nachrichten* paper that he also records his first observation of February 7, 1862, made a week after the Clarks' discovery.

Well before it officially appeared in *Astronomische Nachrichten*, news of the discovery started to percolate through Europe. Jean Chacornac, in Paris, succeeded in observing the companion on the nights of March 20 and 25. It was even observed in early April from the island of Malta in the Mediterranean, by William Lassell who learned of the discovery from the April 1–2 edition of *Galignani's Messenger*, an English language gentlemen's daily of the period published in Paris. As can be expected with a new discovery of this nature, confusion soon ensued. Lassell's observations, with his 48-inch reflector, yielded a location for the new star at considerable variance with those of Bond and Chacornac, leading to doubt that this was actually the long-sought companion. Dr. Peters expressed his initial skepticism that the newly discovered companion was in fact the long-sought dark star in writing to Abbé Moigno, the editor of *Cosmos*, a free-wheeling 19th century astronomical "tabloid". An enigmatically brief account of Peters' views, carried in the pages of *Cosmos* for March 28, 1862, stated "Mr. Peters writes us that he does not accept the identity of the companion that has just been discovered, with the one that is calculated." A week later, however, Peters had second thoughts, and conceded in comments following Chacornac's paper in the pages of the *Astronomische Nachrichten*, that in all likelihood Bessel's companion has indeed been found. The 1862 spring observing season for Sirius was now drawing to a close and there were no further reports of observations for the remainder of the year. In 1862, The French Académie des Sciences, setting aside any doubts concerning the new star, recognized the discovery of the companion of Sirius, and awarded Alvan Clark Jr. the Lalande Prize, and a gold medal, in recognition of the discovery, and the development of the $18\frac{1}{2}$-inch lens. The Lalande Prize is awarded annually in recognition of the year's chief astronomical achievement by the French Académie.

By 1863 the new companion was being observed even as far east as Pulkovo in Russia, where Otto Struve, son of the great Friedrich Wilhelm Struve, began his 15-year series of observations of Sirius in late March. New confusion was contributed that spring by a report in the *Monthly Notices of the Royal Astronomical Society* attributing to M. Gouldschmidt in Paris the observation of six stars in the vicinity of Sirius! In June of 1863 Gouldschmidt, however, protested that he had been misinterpreted and that he did not regard all the "stars" shown in the lithograph entitled "Satellites" of Sirius as true companions of Sirius. By April, Sirius could no longer be easily observed and attention was focused on trying to determine whether or not the companion was actually Bessel's dark star or just an unrelated field star, being noticed for the first time.

The question naturally arises: Why was the companion of Sirius not detected well before the Clarks' accidental discovery in 1862? The famous American double star observer Sherwood Burnham asked this rhetorical question in 1879. At the time of Bessel's discovery of the existence of the perturbed motion of Sirius in 1844, the companion was very near periapse, within 4 arc seconds of Sirius, and certainly could not have been observed. However, by the time it was finally found in 1862, it was 18 years past periapse and 10 arc seconds from Sirius and therefore a relatively easy target. Over the next several years, when observing conditions had not changed by that much, the companion was repeatedly observed by dozens of astronomers and amateurs throughout the U.S. and Europe, some with telescopes with apertures as small as $7\frac{1}{2}$ inches. Thus, there is clear evidence the Sirius' companion could well have been discovered from the mid-1850s on, especially after Peters' work in 1851, which indicated on which side of Sirius the companion should be located.

As mentioned in the last chapter, Urbain J. J. Le Verrier, the man who stunned the world with his prediction for the location of Neptune in 1846, had already, in his capacity as director of the Paris Observatory, personally observed Sirius on, or just before January 27, 1862, at least four days before the Clarks' discovery on January 31. That Le Verrier would actually visit the dome and look through the new telescope would come as a surprise to many then and now. Camille Flammarion, the prolific 19th century astronomer and writer of popular astronomy, who once worked under Le Verrier, doubted that Le Verrier ever even bothered to view his most famous discovery, Neptune, through a telescope. What Le Verrier thought he saw in the glare of Sirius is not clear. But from Chacornac's logs of the 27th, it is evident that Le Verrier asked Chacornac to confirm the detection of something 5 or 6 arc seconds from the star. However, when Chacornac examined Sirius that evening, he found nothing. There the matter rested until March 14, when Chacornac records in his logbook that, Peters, the editor of *Astronomische Nachrichten*, had provided news of the discovery of the companion by the Clarks in Massachusetts. Chacornac looked at Sirius again that evening, finding nothing. Le Verrier, now aware of the discovery, must have pressed Chacornac to try again. Charconac searched for the elusive companion again on the nights of March 18 and 19 but his logs record numerous complaints of difficulties with "rayons" and "vaporeux", the former presumably scattered light from Sirius and the latter mists of the Paris night. Chacornac was finally successful on the early evening of March 20, locating the companion right where Clark said it was. A cautious Le Verrier, however, was reluctant to publicize the observation until it had been confirmed by a second observation on the 25th. Le Verrier presented the news of Chacornac's success to the meeting of the Académie des Sciences on Monday the 24th. Chacornac sent a report to *Astronomische Nachrichten* on the 26th, where it appeared a week after Bond's report. Curiously, there is no mention of Bond or Clark in Chacornac's paper, but he did add an important piece of information. The companion was ten thousand times fainter than Sirius.

At the time of the discovery of Sirius' companion, Le Verrier was deeply involved in an effort to account for an unexplained advance in the perihelion of Mercury's orbit (Chapter 4), which he had discovered eight years earlier. Le Verrier's favored explanation for Mercury's anomalous motion was the existence of the putative planet, Vulcan, believed to orbit between the sun and Mercury, and he was busy orchestrating a search for the phantom planet. In fact there was a flurry of excitement on March 20, 1862, the very day of Chacornac's detection of Sirius' companion, when a sighting of "Vulcan" transiting the face of the sun was claimed by a railway station master and amateur astronomer in England. Bessel's fabled dark stars orbiting Sirius and Procyon were the prime examples of a gravitational anomaly beyond the solar system. The discovery of the cause of Sirius' and Procyon's perturbed motion would have capped Le Verrier's efforts to banish all major remaining unexplained aspects of Newton's Theory of Universal Gravitation. Indeed, Chacornac's logs of March 25 indicate that Le Verrier also urged him to search Procyon for a companion, but he found nothing.

One apparent concrete result of Le Verrier's failure to discover the companion of Sirius was that he finally became convinced that Paris was perhaps not an ideal location for a major observatory. Le Verrier had serious concerns about the suitability of Paris as a sight for a major telescope. Even before he realized that he had been preempted by Clark's discovery, Le Verrier was investigating moving Foucault's telescope. Chacornac's logs contain an intriguing account of Le Verrier returning from a March 9 dinner with Emperor Napoleon III, at which relocating the telescope to Marseille was discussed. It is also evident from contemporary accounts that the initial failure to discover the companion of Sirius led Le Verrier finally to push the French Government to establish a southern observing station of the Paris Observatory in Marseille. Foucault's telescope was moved there in 1864, where it served for the next century, being used for double star work, including observations of Sirius' companion, up until 1962. It is not known what role, if any, Le Verrier played in the award of the 1862 Lalande Prize to Clark. Le Verrier was not a member of the five-member Lalande Committee, which contained three mathematicians and two observational astronomers, associated with the Paris Observatory.

During the American Civil War, the Alvan Clark firm produced spyglasses and, later, field glasses for the United States Navy. After the war there were additional contracts from the Navy for the Clarks to refinish the optics for several telescopes at the U.S. Naval Observatory. As interest in telescopes and astronomy revived in post-war America, contracts for new and larger telescopes began coming in. The Clarks were even able to find a home for their now famous $18\frac{1}{2}$-inch objective. In January 1863, a bidding war between Harvard and the Chicago Astronomical Society broke out to acquire the superb lens. George Bond had long had his eye on the lens. As early as 1862, he mentions in the annual Harvard College Observatory report how it might be acquired for "a trifling expense over the cost of the glass itself." In an obvious appeal to the observatory's committee of overseers, Bond offered his hope that "May we not, even now, indulge the hope that at no distant day this beautiful glass, so great in promise, and so unlikely ever to reach its original destination in Mississippi, may be placed in the hands of the Cambridge observers?" In the end, a well-heeled delegation from Chicago secured the lens. When they arrived at the Clark factory they saw the lens already packed in a box ready for shipping and realized they needed to move fast. They learned that Alvan Clark Sr. had left earlier for Boston to negotiate a sale with

Figure 5.2. The tube of the Clark $18\frac{1}{2}$-inch refractor now on display at the Adler Planetarium in Chicago (permission Adler Planetarium & Astronomy Museum).

George Bond and so they raced after him, cash in hand, and closed the deal with a $1500 down payment. The lens, which was purchased for a total of $11,187, saw many years of active service in Chicago, even surviving the Chicago Fire of 1871. It was eventually housed at the Dearborn Observatory, under the directorship of Thomas Henry Safford, and continued to be used well into the 20th century for double star work, including numerous observations of Sirius' faint companion during the 19th century. The telescope was rebuilt in 1911, and the original wooden tube replaced and eventually went on display at Chicago's Adler Planetarium (Figure 5.2). The famous lens, in a modern mounting, is now part of a telescope still used for scientific research and for public viewing nights at the Dearborn Observatory on the Northwestern University Campus, in Evanston, Illinois.

Throughout the remainder of the 19th century Alvan Clark & Sons successively built an ever-larger series of lenses. Beginning with the $18\frac{1}{2}$-inch objective they repeatedly claimed the title of world's largest refractor a total of five times. A rival 25-inch telescope had been built in England in 1870, but this was soon eclipsed by the construction of the 26-inch U.S. Naval Observatory refractor, completed in 1873. The United States Congress had passed an appropriation to build the new telescope but Alvan Clark, suspicious of post-war government commitments, demanded payment in gold before he would begin. It was only when the industrialist and inventor of the mechanical reaper, Leander McCormick, became interested in underwriting a telescope of a similar size, that a deal was struck whereby Clark would receive payment in gold from McCormick for the first telescope destined for the U.S. Naval Observatory and a twin would be built for the new Leander McCormick Observatory in Virginia. The public, who eagerly followed the progress in the popular press of the new national telescope in Washington, D.C., were soon rewarded when Asaph Hall used it to discover the Martian Moons, Deimos and Phobos, in 1877. In 1879 Otto Struve (Friedrich Struve's son) visited the Clarks and contracted with them for a 30-inch lens for the Pulkovo Observatory in Russia. The subsequent lens turned out so successfully that Czar Alexander III awarded Alvan Clark a gold medal. The next great refractor to be built encountered much greater difficulty.

The San Francisco entrepreneur James Lick left money to build the world's largest refractor on Mt. Hamilton, south of the San Francisco Bay. It required $3\frac{1}{2}$ years and 19 attempts, plus a visit by Alvan G. Clark to the glassworks in Paris, to successfully cast the crown glass blank for the Lick Observatory 36-inch telescope. The Clarks also produced a photographic corrector lens for the 36-inch, which Alvan Jr. traveled to California to install in 1887. The Lick refractor proved enormously productive in the hands of skilled observers such as Edward Holden, James Keeler, and E. E. Barnard. The supreme achievement of Alvan Clark & Sons was the 40-inch Yerkes Observatory telescope completed in 1896 (Figure 5.3). The Yerkes, Lick, and Naval Observatory telescopes are still being used for scientific research today. These monumental instruments and numerous others were widely acknowledged as the finest telescopes in the world and were largely responsible for promoting observational astronomy in the United States to a level that exceeded that of Europe by the 1880s.

Most of the large telescope objective lenses built by the Clarks were of their own distinctive design. Typically, they consisted of an outer doubly convex lens of less refractive crown glass and an inner more refractive lens of flint glass, the latter having a single concave exterior surface and a nearly flat interior surface. The unique feature was the unusually large gap between these two lenses—for example, the Lick 36-inch telescope featured a gap of $6\frac{1}{2}$ inches. The mounting cell, which housed the lenses, was perforated to allow for the free circulation of air to establish thermal equilibrium and easy disassembly for cleaning.

Although the Clarks left no records of their techniques, over the years many visitors, both casual and professional, toured the Cambridgeport workshops and

Figure 5.3. The Yerkes 40-inch telescope on display at the 1893 Columbian Exposition in Chicago, prior to its installation at Yerkes Observatory (photographer, E. E. Barnard, Yerkes Observatory photograph).

provided first-hand descriptions of how the world's largest and finest telescope lenses were produced. The preeminent 19th century American astronomer Simon Newcomb gives an interesting account of the techniques used by the Clarks to grind, figure, and polish the 26-inch doublet destined for the U.S. Naval Observatory. The large disks of flint and crown glass were ordered from a firm in Birmingham, England, since glass of the proper quality was not being produced in America at that time. The raw blanks were first carefully inspected for any flaws such as air bubbles, inclusions, striations, and other inhomogeneities. The tests for inhomogeneities due to annealing stress involved the use of polarized light (Appendix A). Polarized light is made of light waves, whose electric field component is predominantly vibrating in a particular plane. It is the polarization of scattered sunlight from air molecules that causes the intensity of the sky's brightness to change with direction when viewed though polarized sunglasses. The large unfigured crown glass blank was taken outside into the sunlight and placed face down on a black cloth. The lens was then viewed in naturally polarized sunlight, scattered by the earth's atmosphere, through a Nicol prism. Small shadowy interference fringes were noted in the glass blank corresponding to inhomogeneities. These imperfections, however, were deemed not serious and subsequently disappeared during figuring of the lens.

Rough figuring of the spherical surfaces was done using large cast iron tools with the desired shape. The tool was placed on a horizontal turntable, which could be rotated by steam power. The initial cutting was accomplished by emery or cast iron sand grit and required about two weeks. Once a rough figure was established, a pitch lap was used with water and jeweler's rouge to polish the surface. The lens was held on the rotating lap while two workmen imparted a reciprocal motion. In this way, the polishing achieved a high degree of spherical symmetry. It was after this stage that the Clark's artistry came into play. The figured and polished lens had approximately the correct shape but still contained much unevenness and many surface imperfections. The technique which the Clarks used to perfect the lenses was largely self-developed and referred to by Alvan Clark as "local correction". The basic meaning of this term was that rather than working the lens to a precise predetermined mathematical shape, the final figure was achieved by an exhaustive process of testing, figuring, and retesting the lens to remove the last imperfections. Ultimately, through this process they arrived at a figure that was appropriate to the particular pieces of glass and that provided a well-focused image over the desired range of colors. The final high spots were removed, literally, by hand. Alvan Clark, using his bare thumbs coated with damp rouge, applied the final corrections. He maintained that his thumbs were more sensitive and precise than any tool and that he could "feel" the imperfections. Visitors reported that his thumbs had developed permanent deep cracks from this practice.

In order to support this complex process the Clarks built a long tunnel beneath the workshop that could accommodate a large lens. At one end of the tunnel they placed an array of artificial stars and at the other end they had a work area where the focused images could be examined. They also tested the lenses on real stars using a large tube mounted on a tall upright pier. Visitors describe the incongruous appearance of what looked like a giant artillery piece looming above the neighboring rooftops, as if to defend Boston from naval attack.

Alvan Clark died in 1887, at the age of 83, during the final phases of the completion of the Lick telescope. In his later years more and more of the optical testing fell to Alvan Jr. However, in his 81st year the father resumed the painting of portraits, which he had abandoned twenty-five years earlier, completing fine portraits of his sons and grandson. Although he managed only a grade school education, during his long career he earned honorary degrees from several universities, he received a gold medal from Czar Alexander III and the Rumford Medal from the American Academy of Arts and Sciences, all in recognition of his remarkable achievements in optics. In 1879 he penned a brief autobiography at the urging of an acquaintance. His essential practicality and matter-of-fact New England nature is perhaps best expressed in his own terse words as he summed up his life:

> "I have received the degree of A. M. from Amherst, Chicago, Princeton and Harvard. I have read much popular astronomy, but in its mathematics I am lamentably deficient. You will see by the printed papers I send you I have made some use of telescopes." "I will add further what may be of interest. I have always voted with the Republicans when voting at all since they came to power, but have never attended caucuses or held an office. I have never been a church member, nor had either of my parents, but my faith in the universality of God's providence is entire and unswerving. My grandfathers died one at age 87 and the other at 88. I knew them well, and they were good men. Both had been engaged in killing whales.—I have never heard of one of my progenitors—Thomas Clark of the Mayflower was one—as being a bankrupt, or grossly intemperate.—I was never but once sued, and in that case employed Joel Giles as council, who made a compromise without going to trial. I never sued but one man, and that was Collector Austin, and I gained my case. I never studied music or attended an opera in my life, and know nothing of chess or card playing.—I never learned to dance, but was a good swimmer, though lacking generally in the points which go to make an expert gymnast."

Alvan Clark Jr. continued making lenses until his death in 1897. It was Alvan Jr. who was primarily responsible for undertaking the construction of the Yerkes 40-inch refractor in 1893 (Figure 5.3). Alvan Graham remained firmly convinced that the future of observational astronomy lay with the development of larger and larger refracting telescopes. In 1893 he laid out this vision in several papers. He saw no real limitations to the sizes that lenses of the future might attain, but remained pessimistic about the possibility of improving reflecting telescopes. Others such as George Hale of the University of Chicago, who had put together the funding and support for the 40-inch telescope at Yerkes Observatory, in Williams Bay, Wisconsin, were beginning to see things differently. It was Hale who would later go on to build the large reflectors at Mt. Wilson and Mt. Palomar in California. However, Clark thought a 50-inch refractor was possible and his firm already possessed the facilities to produce a lens of that size. Such a behemoth, which would have had a length of nearly ninety feet, never came to pass and the Yerkes telescope remains to this day the world's largest such instrument. A 41-inch objective was attempted by the Parsons firm of Newcastle, England, for the Russian Observatory at Simeis in the Crimea in 1926. It failed testing and was abandoned in favor of a similar-sized reflecting telescope.

The large refractors built by the Clarks became the centerpiece of many major observatories, principally in the United States but also in Europe, during the latter half of the 19th century. These instruments were a key factor in advancing astronomy in America and were responsible for a number of well-known discoveries made during that period. Examples include E. E. Barnard's 1892 discovery of the first new moon of Jupiter since the days of Galileo, Almathea, and James Keeler's observation of a new gap in the outer A ring of Saturn in 1888. Both these discoveries were made with the Lick Observatory 36-inch telescope. An ironic parallel to the discovery of the companion of Sirius occurred early in the morning of November 15, 1896. Professor J. M. Schaeberle was examining the star Procyon with the Lick 36-inch telescope when he observed a faint, but clearly defined, yellowish 11th magnitude star bathed in the glare of Procyon, some 4.35 arc seconds away. Schaeberle had finally succeeded in observing Bessel's second dark star. The Procyon system, like that of Sirius turns out to harbor a very faint and very special companion star in a highly eccentric orbit with a 40.65-year period. Perhaps the most famous observations, however, were Percival Lowell's enchanting descriptions of Mars towards the end of the century at the Lowell Observatory in Flagstaff, Arizona. Lowell had commissioned the Clark firm to construct the 24-inch lens of his famous telescope.

Alvan Graham Clark was of a more cosmopolitan temperament than his father. He took great delight in literature and could quote long verses of poetry from memory. He enjoyed entertaining visitors to the firm, often sitting for hours with the poet William Longfellow discussing poetry. The Yerkes 40-inch was Alvan Graham Clark's last great project, though the actual work for that lens was largely done by Carl Lundin, the Clark's long-time master optician. Alvan Jr. died shortly after personally supervising the installation of the Yerkes 40-inch lens in 1897. At his funeral in June of 1897, his pallbearers included Edward Pickering, director of the Harvard College Observatory, Asaph Hall of the U.S. Naval Observatory, the discoverer of the moons of Mars and fellow Bostonian Percival Lowell, who was just becoming famous with his observations of Mars.

Alvan Clark and his sons left no progeny with either the interest or inclination to carry on the family business. After the death of Alvan Jr., the firm continued to produce telescopes under the direction of Carl Lundin. It was Lundin who built the 13-inch astrograph lens used by Clyde Tombaugh to discover the planet Pluto in 1930 at the Lowell Observatory in Flagstaff, Arizona. In 1933 the Alvan Clark & Sons firm was sold and became a subsidiary of the Sprague-Hathaway Mfg. Company; which itself went out of business in 1958. A large number of Alvan Clark telescopes still exist, particularly at colleges, universities, observatories, and planetariums throughout the world. Many of these retired instruments are still used for public viewing nights. Many more, smaller Clark telescopes are in private hands where they are highly valued as functional antiques, selling for over five thousand dollars per inch.

Figure 1.4. A tetradrachma from Alexandria from the reign of the Roman Emperor Antoninus Pius (left), showing the Phoenix crowned with a halo (Milne 1737; Courtesy of the Michael R. Molnar Collection).

Figure 2.2. A 3rd century BC coin from the Greek island of Ceos. The reverse (right) shows a dog surrounded by radiant rays (Sear 3079; courtesy of the Michael Molnar R. Collection).

Figure 2.3. Sirius in the form of a rampant wolf-like dog, from the 9th century Codex Vossianus Latinus manuscript in Leiden (Reprinted with permission from *Sky & Telescope*, June 1992).

Figure 2.4. An Alexandrian drachma from the reign of the Roman Emperor Antoninus Pius, showing a representation of the Egyptian goddess Isis astride the Greco-Roman canine representation of Sirius (Milne 2358; courtesy of the Michael R. Molnar Collection).

Figure 9.3. The relative sizes of Sirius B and the Earth (Space Telescope Science Institute).

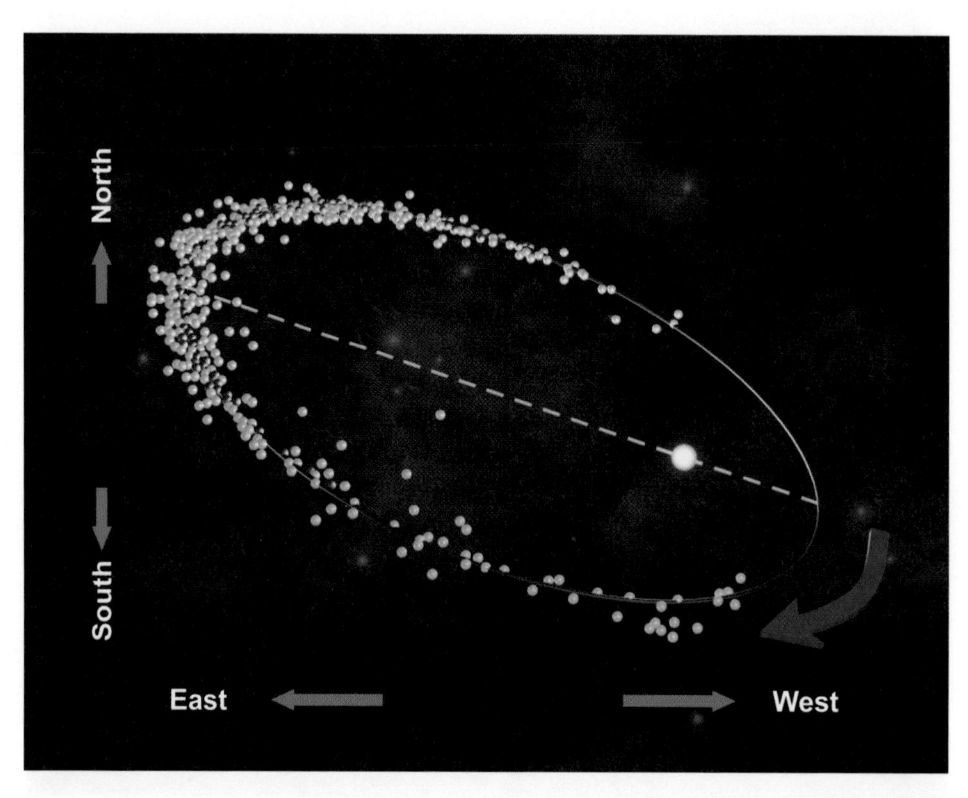

Figure 12.2. The orbit of Sirius B showing the locations of the historical visual (green dots) and photographic observations (orange dots).

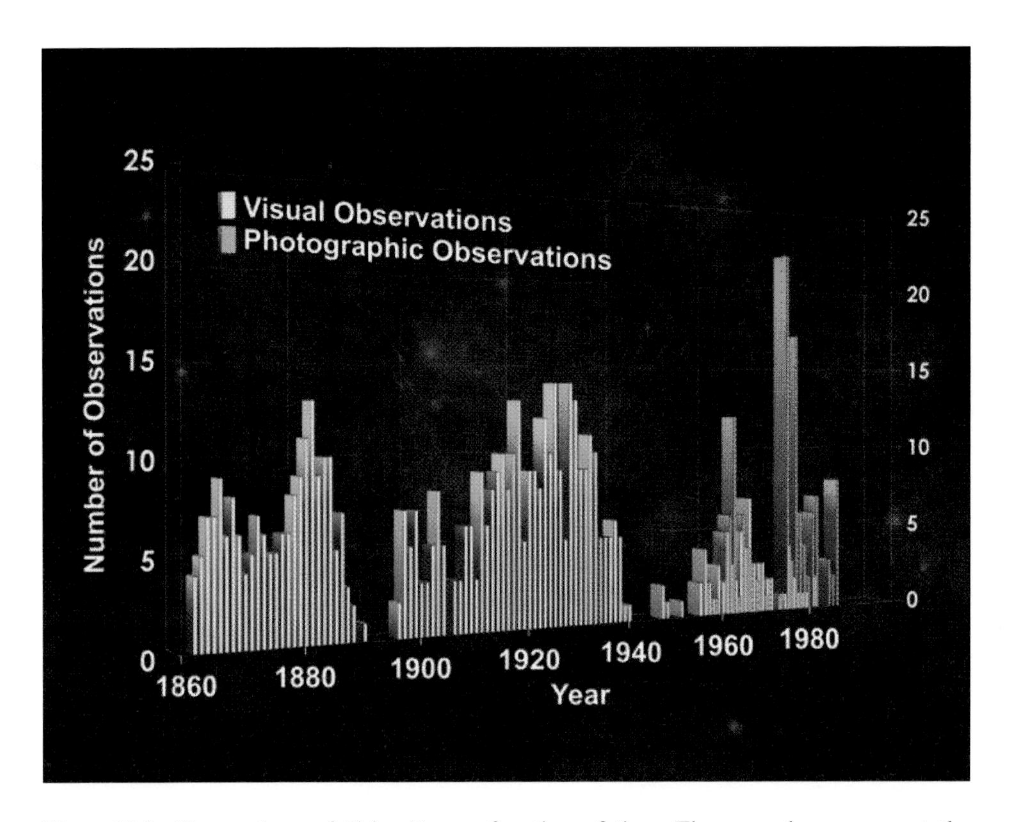

Figure 12.3. Observations of Sirius B as a function of time. The green bars represent the number of visual observations in a given year and the orange are photographic. The gaps near 1896 and 1944 correspond to the period when Sirius B was nearest Sirius and lost in its glare.

Figure 13.1. An artistic rendition of Sirius A and B (Space Telescope Science Institute).

Part Three

The Physics of the Stars

"If we are not content with the dull accumulation of experimental facts,
if we make any deductions or generalizations,
if we seek for any theory to guide us,
some degree of speculation cannot be avoided"
—Sir Arthur Eddington, 1920

The 19th century witnessed enormous advances in our understanding of the sun and the stars. At the beginning of that century, even as distinguished an astronomer as William Herschel could seriously advance the view that the sun was a habitable world whose surface was obscured by thick clouds of radiant brilliance. As the century drew to a close, an entirely new stellar universe had emerged. The sun and the stars were fiery furnaces powered by their own self-gravity and destined to end their relatively brief lives as cold dark lumps of dense matter. This somewhat pessimistic picture of the stars was painted with the help of the new tool of spectroscopy, which split the light of the stars to reveal their chemical compositions and hint at their temperatures. The newly developed laws of thermodynamics appeared to govern their evolution and seal their fates.

As the 20th century dawned, astronomers had accumulated a mountain of spectroscopic data on the stars, but lacked a suitable theory to make sense of it all. The first great advance was the realization that two dramatically different types of stars existed. Familiar stars like the sun might seem large, but they were dwarfed by a population of immense stars, thousands of times larger and more luminous than the sun. Some of these stars had grossly distended atmospheres and average densities many times less than that of air. This realization was shortly followed by the discovery that there also existed an entirely different class of stars with opposite characteristics. These stars were the size of the earth, but had masses comparable with that of the sun. It was the new discoveries of the atom and its components—the electron, proton, and neutron—that eventually explained how nature managed to accommodate all of these totally different stellar types. Many of the discoveries that shaped these new views of the stars directly involved Sirius and its curious companion.

6

An Odd Pair

"As one great Furnace flam'd yet from those flames
No light, but rather darkness visible"
—John Milton, *Paradise Lost*, 1667

In the period immediately after its discovery in 1862, observers in Europe, England, and America continued to report the changing position of the new star, or "Sirius *comes*" (Latin for companion), as it was often called at the time. Among the questions on everyone's mind was: Is this in fact Bessel's Dark Star, and if so why is it so faint? Does it shine by its own light or does it merely reflect light from Sirius? Initially some had suggested the faint companion was a large planet orbiting Sirius and indeed it was frequently referred to as the "satellite" of Sirius, implying that it might not be a star in its own right. At this time the distinction between a large planet and a star was not so clear as to promote the type of sensation that attended the announcement of the discovery of the first extrasolar planets in the present day.

Most observers at the time realized that it would require several years of careful measurement to establish exactly how the new companion moved with respect to Sirius. Would it share Sirius' proper motion or was it simply a faint background star which would fall behind, as Sirius moved to the southwest? Most important of all: Would the companion move in the manner predicted for the mysterious perturbing body?

A prematurely negative answer to questions about the companion's motion came in 1864. In early May of that year a letter to the *Monthly Notices of the Royal Society* arrived from Russia. The author was Otto Struve (1819–1905), who had succeeded his illustrious father, Friedrich, as the director of Pulkovo Observatory near St. Petersburg. In the letter Struve described his observations of the new companion during the years 1863 and 1864. From these, he calculated changes in the azimuthal angle and the separation between the two stars and compared the motion with the predictions of

Figure 6.1. Otto Struve (1865) (photograph in the collection of the late Nils Lindhagen, reproduced by permission of Anne Lindhagen).

Truman Henry Safford. The observations appeared to show that the companion was separating from Sirius in a direction and at a rate expected for an unrelated background star, not sharing Sirius' proper motion. When Struve (Figure 6.1) added in Bond's 1862 discovery observation it also seemed to agree with the conclusion that the two stars were not related. However, Struve shrewdly remained skeptical of his own calculations and limited data: "Nevertheless, I do not yet regard this conclusion as sufficiently established." He reminded astronomers that 70 years earlier William Herschel had examined the part of the sky in which Sirius was then located, and had not noted any star corresponding to the present companion.

At the latitude of St. Petersburg, Russia (61°N), Sirius is very difficult to observe, since it never rises more than 13° above the horizon. This low elevation resulted in very turbulent viewing conditions, which makes precise position and brightness measurements difficult. In fact, at Pulkovo the optimum observing season is restricted to a brief period, each February and March, when the star culminates just after midnight.

In spite of these difficulties Struve had begun systematically observing Sirius and its new companion in early 1863.

By 1866 the resolution of the issue of whether or not the new companion was indeed Bessel's dark star had become clear. Indeed, in May of that year, Simon Newcomb at the U.S. Naval Observatory in Washington, D.C. ironically remarked "If there is any doubt of the identity of the two objects, an appreciable fraction of such doubt must pertain to the question whether the Neptune we see with our telescopes is the same body that disturbs the motion of Uranus."

In 1866 Struve had accumulated sufficient observations to reach three fundamental conclusions about the nature of the faint companion. First, he showed conclusively that the new star was in orbit about Sirius and was not simply a faint background star, whose current position happened to coincide with that of Sirius. He did this by demonstrating that over the prior three years the proper motion of Sirius would have caused the two stars to separate by an amount well in excess of the uncertainty of his observations. Second, he went on to demonstrate that the mutual motion of the two stars was in fact consistent with the newly discovered star being "Bessel's obscure body". These deductions, shrewd as they are, were well within the capabilities of numerous other astronomers of the day, given the same observational data. What is truly remarkable, however, was that Struve in 1866, at a time when astronomers were just beginning to grapple with the physical characteristics of stars, came to a startling conclusion about the nature of Sirius B (or the "satellite" as he called it), namely that it was of a "very different physical constitution" from Sirius A. He stated this conclusion in the final paragraph of a brief letter he wrote to the *Monthly Notices of the Royal Astronomical Society*:

> "Admitting then that the observed satellite is identical with Bessel's obscure body, the above given relation of $d = 3.087\Delta$, [*see below*] indicates that its mass must be estimated approximately half that of *Sirius* itself. If both bodies had the same physical constitution, this relation of the masses would assign to the globe of the satellite a diameter of only 1.26 times smaller than that of the principle body, and therefore, considering the extraordinary brightness of the large star, we would be induced to place also the satellite in the first class of magnitude. With this conclusion the observed brightness of the companion forms a manifest contradiction. It is commonly said to be of the 9th or 10th magnitude; and only in the Spring of 1864 I have noted it once as of 8th magnitude, probably on account of an extraordinary favorable state of the atmosphere. Hence it follows that, to maintain the identity, we must admit that both bodies are of a very different physical constitution. That the light of the satellite is gradually increasing, as I was inclined to suppose from the comparison of my observations of 1864, with those of 1863, has not been confirmed by later observations; but in our latitude the estimation of the brightness depends too much from the condition of the atmosphere to admit of an accurate judgment in this respect."

Pulkova, April 15, 1866

Struve couldn't be clearer. He used the ratio of the observed angular separation of the two stars (d) to the separation of Sirius from its apparent system center of mass (Δ), to determine the ratio of the masses of the two stars to be 2:1 (the modern value is (2.03:1). He then reasoned that if they were made of the same material (i.e., had the same density) then the diameters of the two stars would be in the ratio of 1.26:1. He next went on to compare the brightness of the two stars, noting that he had once estimated the brightness of Sirius B to be 8th magnitude (the modern value is 8.44) on a night of exceptional "seeing". Sirius A is thus 10,000 times brighter than Sirius B, yet is only 25% larger in diameter. Struve rejected this paradox and correctly concluded that the two stars are of a fundamentally different constitution. Just how different were the constitutions of these two stars would not be revealed for another 74 years.

Following the initial discovery of the companion to Sirius, the new star was widely observed and commented on by many observers. One interesting, and in hindsight highly significant, observation was that of George Knott, Esq., an amateur astronomer, who observed Sirius in early 1866 from Woodcroft Observatory, Cuckfield, England. He is the first to comment on the color of the companion. "Happening to turn my $7\frac{1}{2}$-inch Alvan Clark refractor on *Sirius*, on the 24th of January, I was surprised to find the small companion, notwithstanding the bright moonlight, a tolerably conspicuous object. Its colour was a fine pale blue (about Blue[3] of the Late Admiral Smyth's chromatic scale) .... ." No one seemed to notice Knott's remark about the companion's blue color, or if they did they attached little importance to it at the time. It would be another five decades before the world's largest telescope would be used to observe the spectrum of the new companion and confirm the blue–white distribution of its light. These same decades would also witness the widespread application of spectroscopy to the task of classifying the kingdom of the stars.

Spectra are formed by spreading a beam of light into its constituent wavelengths or colors. The most familiar method of doing this is to use a simple glass prism to display the colors of the rainbow. This is the technique used by Isaac Newton in 1665, who first showed that white light is a combination of different colors. A more precise technique, now widely used in astronomy, is to allow light to be reflected (or transmitted) by a glass plate which contains thousands of precisely inscribed parallel and equally spaced lines or grooves. Such a device is called a diffraction grating and uses the interference properties of light waves to produce a spectrum. The familiar rainbow pattern of reflected colors seen when CD disks are viewed at an angle is due to the diffraction of light from the uniformly spaced circular pattern of tiny pits that encode the information on the disk. In order to make a detailed study of the light from the stars it is necessary to precisely control how the light falls on a prism or grating by tightly constricting the angle of the incoming beam.

In 1814, Joseph Fraunhofer (1787–1826), a brilliant Bavarian optician and instrument maker in Munich, noticed a series of odd dark lines in the image of a thin narrow shaft of sunlight that he had passed though a glass prism and then viewed with the small telescope of a surveyor's theodolite. In the band of dispersed sunlight, or spectrum, he saw a host of narrow dark lines, impressed upon the rainbow of colors produced by the prism. After conducting numerous tests and experiments Fraunhofer concluded that the dark lines were intrinsic to the sunlight and did not originate within his instrument. The dark lines that Fraunhofer saw corresponded to multiple images, each at a different wavelength, of the slit or the narrow opening that defined the beam of sunlight falling on the prism. The relative darkness of the lines indicated that the light of certain wavelengths seemed to be absent. Physicists were already familiar with bright lines seen in the spectra of laboratory flames and sparks, when observed under similar circumstances. Fraunhofer mapped the dark lines in the solar spectrum, recording more than 570, which have become known as "Fraunhofer lines". To the most prominent of these lines he assigned letters of the alphabet. Fraunhofer also noticed that what he termed the prominent "D" line seemed to coincide with the position of a conspicuously bright yellow line commonly seen in the laboratory spectra of flames. This line later became associated with the element sodium. This coincidence would later prove to be a significant clue as to the nature of starlight.

Fraunhofer next turned his spectroscope to Venus and Sirius. Since these objects are effectively point sources of light, no slits were necessary to define the light beam. The spectra of these sources, however, were much fainter than that of the sun and consequently difficult to view. Nevertheless, Fraunhofer noted that the spectrum of Venus appeared similar to the sun, while that of Sirius was clearly much different. The spectrum of Sirius exhibited only three lines, one in the green and two in the blue. At the time he could not clearly identify these lines with any of those he had seen in the sun. He also viewed the spectra of a few other 1st magnitude stars but only wrote that they all seemed to differ. Nine years later, Fraunhofer attached a similar spectroscope to an astronomical telescope and again viewed the dispersed spectra of a handful of bright stars, including Sirius. While the spectra of some stars seemed to resemble what he had observed in the sun, others including Sirius were markedly different. Sirius again only showed the previously seen three strong dark lines, but this time Fraunhofer could associate them with similar dark lines in the solar spectrum such as the "F" line, now known to be due to the element hydrogen.

Because of the difficulty of collecting enough light from a star to view its spectrum, physicists and astronomers had initially concentrated on the much brighter and easier to observe sun. Slowly, some of the lines in the solar spectrum were recognized as corresponding to the bright lines produced in laboratory flames, which were known to be associated with common chemical elements such as sodium. The real meaning of the solar Fraunhofer lines, however, remained largely a mystery. All of this changed abruptly in 1857 when the Austrian physicist Gustav Kirchhoff (1824–1887) and the chemist Robert Bunsen (1811–1899) published a series of papers summarizing their exhaustive study of the spectra of various laboratory flames and the solar spectrum. They demonstrated hundreds of precise correspondences between the bright lines produced in laboratory flames, to which were added various chemical elements such as sodium, calcium, and iron, and the Fraunhofer lines of the solar spectrum. It was suddenly obvious that the atmosphere of the sun contained many of the same elements which chemists had identified here on earth and that the signature of these elements could be recognized by the unique sets of emission or absorption lines they produced.

Kirchhoff went further, providing the explanation for why dark lines and not bright lines were seen in the solar spectrum, by establishing three principles now known as "Kirchhoff's laws of radiation". These laws state, first, that the light produced by a hot glowing solid or dense gas is a continuum, having no bright or dark lines. Second, however, if the source is a hot thin diffuse gas, like that in a flame, then bright lines are produced, which are distinct for each chemical element. The third and final law states that if a source of continuum radiation, such as a hot solid or dense gas, is viewed through an intervening cool gas, then dark lines will be formed which are characteristic of the elements in the cool gas. It was now clear that the dark lines in the solar spectrum were being formed in the solar atmosphere when light from underlying hotter layers passes through upper cooler layers where light of specific wavelengths was being selectively absorbed out. Suddenly the light from the sun and the stars acquired an entirely new meaning.

One year after the first sighting of Sirius' companion, this new tool of spectroscopy seemed to suddenly explode into prominence among stellar astronomers. It is spectroscopy that would prove to be the key to understanding the nature of the stars and to solving the mystery of Sirius' small faint companion. The watershed year was 1863 and a trio of amateur and professional astronomers focused on stellar spectroscopy. In New York City, Lewis M. Rutherfurd (1816–1892), a wealthy lawyer turned amateur astronomer, built a spectrograph and observed the spectra of a number of stars. Rutherfurd, it will be recalled was the second person to confirm the companion of Sirius in March of 1862. He was also a very talented instrument maker and early practitioner of astronomical photography. Rutherfurd not only reported what he viewed through his spectrograph but he noted that stellar spectra could be grouped into three general classes based on their appearance. In one class he placed "golden and red" stars, such as the sun, with numerous dark lines. In the second class were "white" stars such as Sirius, which were clearly unlike the sun, exhibiting only few strong lines. The final class consisted of white stars that showed no clear lines in his instrument.

William Huggins (1824–1910), like Rutherfurd, was also an independently wealthy amateur astronomer who acquired an early interest in spectroscopy and photography. However, unlike Rutherfurd, Huggins had no university training in the physical sciences. Beginning in 1856, Huggins built an impressive private observatory in his household garden on the outskirts in London. Tiring of standard astronomical observing he decided to take up spectroscopy after attending a popular lecture on Kirchhoff's "great discovery" concerning the solar spectrum. Huggins resolved to apply these new methods to the study of the stars. His first 1863 paper on stellar spectroscopy included a drawing of the spectra of Sirius and three other stars. Not content to simply describe stellar spectra, Huggins and his neighbor, William Miller, measured the positions of stellar lines with a micrometer and compared these positions with solar Fraunhofer lines and the lines seen in the electric spark spectra of various chemical elements. In this way the two were able to identify lines due to hydrogen, sodium, and magnesium in the spectrum of Sirius, with hydrogen being the most prominent. In 1880, using the new, more sensitive dry photographic plates, Huggins obtained the first extensive photographic spectra of many stars, including Sirius. Much of Huggins' later work focused on the study of the chemical content of stellar spectra.

In Rome, Father Angelo Secchi (1818–1878), a Jesuit priest and astronomer working at the Collegio Romano, also began publishing his observations of stellar spectra in 1863. Initially Secchi classified stellar spectra into two classes, red and yellow stars, like the sun, and blue–white stars, like Sirius. Later Secchi, like Rutherfurd, expanded his spectra into three classes. Class I consisted of blue stars, like Sirius, which show a small number of hydrogen absorption lines and not much else. Class II consisted of yellowish stars, similar to the sun, with spectra rich in narrow dark lines. Class III stars were reddish like Betelgeuse and were dominated by broad dark bands. Later Secchi added a Class IV for stars where strong carbon lines were seen. Up until his death in 1878, Secchi visually classified the spectra of over 400 stars, far more than anyone else.

Sirius, being the brightest star, was a frequent target of these early investigations and thus served as a prototype, along with the sun, for early attempts to classify stellar spectra. All three of these early pioneers initially worked visually sketching the spectra of various stars and trying to classify the spectra based on their appearance. Although these classification schemes differed, they all recognized the distinctive stars of the "Sirian" type with their strong blue spectra, punctuated with several prominent absorption lines. In most instances these schemes involved either implicit or explicit assumptions that the various classes reflected a sequence based on stellar evolution. Huggins on the other hand, with his focus on chemical content, viewed the differing spectra as a consequence of strong compositional differences in stellar atmospheres. Between 1863 and about 1890 several more elaborate systems of classifying stellar spectra were developed, the most popular of which was due to Hermann Carl Vogel (1841–1907) in Germany, which focused on an explicit scheme of stellar evolution.

In 1885 a new contender using improved techniques entered the debate concerning stellar classification. Edward C. Pickering (1846–1919), a Massachusetts Institute of Technology physicist, had been hired by the Harvard College Observatory in 1877 to become the next director. Pickering recognized the importance of hard observational data, in particular data that systematically recorded the brightnesses and the spectra of the stars. In pursuit of this objective he organized a series of programs at Harvard to collect previously unimaginable quantities of stellar spectra. First, however, he needed money. In 1882 the widow of Henry Draper, a well-known New York amateur astronomer and pioneer in photography, approached Pickering about setting up a fund to support a program of stellar spectroscopy at the Observatory to memorialize her husband. Henry Draper had been the first person to succeed in photographing the spectrum of a star (Vega) in 1877. Pickering used the newly established Henry Draper Memorial Fund to initiate a unique program that produced stellar spectra on a truly industrial scale. The new technique he employed was that of an objective prism, in which the telescope aperture is covered by a thin wedge of glass forming a shallow angle prism. The resulting focused stellar images were no longer points of light, but tiny dispersed spectra. Pickering also used a wide field of view of up to 10 degrees, so that the spectra of dozens of the brighter stars in the field could be obtained simultaneously. His chief innovation, however, was to record these images on 20 × 25 cm photographic plates. In this way the spectra could be classified by measuring the photographic plates in the comfort of an office. An initial program of objective prism spectroscopy was begun in 1885 and completed in January 1889, covering the northern sky and extending down to declinations of −25 degrees.

One problem Pickering faced was what to do with this wealth of data. His solution was to assign Williamina P. Fleming (1857–1911) the task of classifying the spectra. Fleming, who had no formal astronomical training, had been employed by the Observatory to perform routine mathematical computations and copying. Fleming it turns out was a brilliant choice, since she brought few astronomical preconceptions about the nature or meaning of the spectra to the task of spectral classification. This turned out to be an advantage, since much of the early thinking about stellar classification was based on what are now known to be incorrect assumptions about how stars evolved. She began with Secchi's four basic spectral types and proceeded to elaborate on them using the letters A, B, C, D, E, F, G, H, I, K, L, M, and N. This system was to become the forerunner of the modern notation for stellar spectral classifications. Stars of type A and B were bright blue stars like Sirius and those in Orion, while stars of type F and G had spectra similar to that of the sun, and K and M stars were characterized by reddish spectra containing strong dark bands, in addition to the narrow metallic lines seen in other stars. Later, in 1897, after the discovery of the element helium in the solar spectrum, an additional spectral type "O" was introduced to accommodate the hot bright blue stars in Orion, where the lines of ionized helium are particularly strong.

Along with the task of classification Fleming also estimated the photographic magnitudes of the stars by measuring the density of dark stellar images on the plates in the vicinity of Fraunhofer's "G" line in the blue–green part of the spectrum. Unlike the earlier classification schemes, no attempt was made to identify the different classes of stars with differing stages of stellar evolution. Nevertheless, most astronomers clearly viewed the Harvard spectral classes as representing a sequence of stars having different temperatures. The final catalog of spectral types and magnitudes for some 10,331 stars was published in 1891 as *The Draper Catalog*. Sirius had catalog number 3797, and its spectral type listed as an ambiguous "A?" since Sirius proved too bright on the photographic plates and its spectrum was overexposed.

Pickering followed *The Draper Catalog* with a more detailed classification of the spectral types of the brighter stars using higher dispersion spectra. He entrusted the classifications of these spectra to Antonia C. Maury (1866–1952), who was a niece of Henry Draper and who had been a research student at Harvard. Maury, faced with more detailed spectra, came up with a much more elaborate classification scheme involving some 23 types. Her most important contribution, however, was to recognize a distinction between stars which showed a similar complement of spectral lines, but for which the detailed appearance of these lines differed. To accommodate this distinction she created subclasses based on the appearance of the lines. Stars with normal, fairly distinct lines, like the sun (the majority) were labeled "a" stars, while those with wide and hazy appearing lines were labeled "b", and stars with sharp narrow lines were labeled "c". The difference between subclasses "a" and "b" and the subclass "c" stars would turn out to be an important clue to the luminosity of different stars. At the time Maury's study was published in 1897, however, her subclasses were viewed by many, including Pickering, as a nonessential complicating factor in achieving a uniform scheme of stellar classification.

In 1896 Pickering hired Annie Jump Cannon (1863–1941) as an assistant at the Harvard College Observatory. She had obtained graduate level training in astronomy from Radcliff College and Wellesley College, and Pickering asked her to examine the spectra of some of the peculiar stars found in the Henry Draper survey. In 1901 she had revised the Maury classifications by abandoning the subclasses, based on line appearance, and going back to the system of Fleming. However, Cannon also greatly simplified the older classes by using only the letters O, B, A, F, G, K, M and adding decimal subdivisions to the letters to accommodate intermediate classifications. For example, a star showing hydrogen line strengths mid-way between A and F would become an A5 star. At an international meeting in Bonn in 1910 the basic Harvard scheme, with some modifications, was adopted as the standard system for classifying stellar spectra. This was in large part due to the effective lobbying for the system by Pickering and to the fact that Harvard had already classified most of the brighter stars in both hemispheres.

The Harvard classification system of Cannon, although based on the relative strengths of various key spectral lines, was supposed to be free from considerations of stellar evolution. It was widely recognized, however, that the system did in fact represent a relative temperature sequence with the blue O, B, and A stars as the hottest, F and G stars at intermediate temperatures, and the reddish K and M stars at the cool end. Since temperature could well follow an evolutionary sequence, relative stellar ages became permanently associated with the letters of the Harvard scheme. In the minds of many the Harvard spectral classes represented a cooling sequence, with the blue stars being the youngest. Although this interpretation is no longer valid, astronomers habitually refer to "early-type stars" when discussing O, B, and A stars and "late-type stars" when speaking of K and M stars. As will be seen in the next chapter, it is mass, not simply temperature, that is the basic organizing parameter of the Harvard system.

In 1912 Pickering initiated an even grander program which would include the classification of nearly all stars of magnitudes brighter than 8 or 9 in both hemispheres. This was an immense undertaking and Annie Cannon, with an average female staff of just five, spent the next four years carefully measuring thousands of photographic plates and classifying over a quarter of a million stars (Figure 6.2). Cannon herself was responsible for determining the appropriate spectral class of each star. The result was a huge catalog of 224,300 stars, covering the entire sky. Sirius was simply listed in the *Henry Draper Catalog* under HD 48915 ("HD" for Henry Draper Catalog), as an "A0" star with a photovisual magnitude of −1.58, along with its coordinates for the year 1900. The catalog filled nine volumes of the *Annals of the Harvard College Observatory* and appeared between the years 1918 to 1924. To early 20th century astronomers studying the stars, the *Henry Draper Catalog* was the astronomical equivalent of the *Oxford English Dictionary*, whose serial publication was unfolding during the same period.

While astronomers were busy trying to classify stellar spectra they were also beginning to pay serious attention to another aspect of starlight: the information that

Figure 6.2. Edward Pickering in 1913 with the female staff responsible for the production of the *Henry Draper Catalog*. Annie Jump Cannon is second from the right from Pickering in the middle row (permission Harvard College Observatory).

it conveyed about the velocity of the stars. The radial velocity, or the component of velocity directly along the line of sight, of a star can be measured using the Doppler shift. The name Doppler is due to Christian Doppler (1803–1853), a mathematics professor in Prague, who in 1842 discovered the principle that bears his name through the study of sound waves, and who speculated that a similar effect could be observed with light waves. Astronomical Doppler shifts are changes in the wavelengths of light that are proportional to the relative velocity between a source and the observer. Shifts are to longer wavelengths—i.e., the red—when the velocity is positive and the distance between source and observer is increasing, and are to the blue or negative when the source and observer are approaching each other. Some, including Doppler himself, thought this effect actually explained the different colors of the stars. The Austrian physicist Ernst Mach (1838–1916) finally pointed out in 1860 that the magnitudes of expected Doppler shifts were small and would involve slight displacements of the lines that would require very careful measurement, if there was any hope of success in determining stellar velocities. This meant that the stellar spectrum must be spread out as much as possible to detect the small wavelength shifts. Spreading out the light has the effect of creating a fainter spectrum relative to a less dispersed spectrum. Spreading the light with the early spectrographs also meant adding more prisms to bend the light through a larger angle entailing additional light loss with each successive prism. Some of the early designs bent the light path through nearly 360 degrees to achieve the desired amount of dispersion. Finally, the small displacements had to be estimated by eye, making matters even more difficult.

Observations of stellar Doppler shifts had a long and initially unfruitful gestation period, during which the radial velocity of Sirius was a prime subject of contention. As early as 1868 Father Angelo Secchi of the Vatican Observatory attempted to measure the radial velocities of bright stars with spectral types similar to Sirius. He used the Fraunhofer F line (the Balmer β line of hydrogen) as a fiducial marker to set his micrometer cross hairs and then observed the same line in other stars. No Doppler shifts were observed and Secchi concluded that none of the stars had a velocity that differed from that of Sirius by more than 150 to 180 km s^{-1}. Just a few months later William Huggins reported to the Royal Society that he had also observed the F line in Sirius, but had compared its position with that of the F line from a Geissler tube (an early gas discharge tube) containing hydrogen. Huggins measured a Doppler velocity of +47 km s^{-1} for Sirius. This prompted Secchi to reobserve Sirius and also to compare the position of its F line with an F line generated in the laboratory. He noticed a small displacement and stated that "this could imply a very rapid motion of the star." A spirited dispute then arose between Huggins and Secchi about the priority of the discovery of stellar Doppler shifts and the reality of the measured velocities of Sirius. Several programs devoted to the measurement of stellar radial velocities were initiated at major observatories over the next two decades. Eventually doubt was cast over many of these early visual measurements when, in 1890, the results of the Greenwich Observatory program were published and the six measurements of Sirius' Doppler velocity were seen to range from +12 km s^{-1} to −79 km s^{-1}! The chaos was eventually resolved in the late 1880s when Herman Carl Vogel, and his colleague Scheiner at Potsdam, refined the spectrograph to systematically remove systematic errors, and to replace the human eye with photographic plates that permanently recorded the position of stellar lines and reduced the effect of judgment in measuring small displacements. Their results for the Doppler velocity of Sirius finally settled on an average value of −15.6 km s^{-1}.

It wasn't until 1896 that truly reliable photographic stellar Doppler shifts began to be systematically measured and the radial velocity of Sirius finally converged on small values, between −1 km s^{-1} and −5 km s^{-1}. These results were first obtained at the Lick Observatory and analyzed by Professor William Wallace Campbell (1862–1938). In 1905 Campbell published his results, along with those of other observers, and was able to actually detect the expected change in the radial velocity of Sirius A due to the orbital motion of Sirius B. Campbell used the opportunity of the changing Doppler velocity of Sirius to work out a consistent theory of how to combine measurements of the visual orbit with measurements of the radial velocity to fully specify the orbit of a binary star. In the case of Sirius, where the visual orbit was relatively well known, Campbell could, for the first time, resolve the ambiguity with respect to the direction of Sirius' binary orbit. Did the two stars orbit counter-clockwise or clockwise? An ambiguity about the directional sense of the orbit arises if only positional measurements are available, because a counter-clockwise orbiting star is indistinguishable viewed from north of its orbital plane from a clockwise orbiting star viewed from the complementary angle, south of its orbital plane. Campbell found that Sirius orbits counter-clockwise and that we view it from below its orbital plane. He was also able to measure the constant radial velocity of (the center of mass of) the Sirius system with

Figure 6.3. John Ellard Gore (permission AIP Emilio Segré Visual Archives).

respect to the sun, finding this velocity to be $-7.36\,\mathrm{km\,s^{-1}}$, meaning that Sirius is approaching the sun at this rate. The modern value is $-8.54\,\mathrm{km\,s^{-1}}$. The binary motion of Sirius only produces a net orbital variation of some $\pm 10\,\mathrm{km\,s^{-1}}$ in the radial velocity of Sirius A during the course of its fifty-year period.

The obvious paradox first raised by Struve in 1866, that although Sirius and its companion have similar masses yet exhibit a vast disparity in brightness, went largely unnoticed throughout the 19th century. There was, however, one person who in 1898 independently came to very similar conclusions to those of Otto Struve thirty years earlier. Mr. John Ellard Gore (Figure 6.3) was an active Anglo-Irish amateur astronomer, and popularizer of astronomy in late Victorian England. Gore (1845–1910) was born in Ireland and had studied engineering at Trinity College in Dublin. He served as a public works engineer in the Indian Punjab for eleven years and returned to Ireland in 1877, where he lived off his Indian service pension. Gore, who appears to have had no formal training in astronomy, began to study the sky while in India. Later in Ireland, he used a small three-inch telescope and binoculars. With these simple tools he made numerous contributions to the *Monthly Notices of the Royal Astronomical Society* on variable and double stars. He was a fellow of the Royal Astronomical Society and the Royal Irish Academy and a founding member of the British Astronomical Association. He was also a very successful writer of popular astronomy books in the latter part of the 19th century. One of the more widely read books was *Astronomy*, which he helped co-author with Agnes Clerke, another active Anglo-Irish popular writer of the era.

In 1891 Gore published a short note in the *Journal of the British Astronomical Association*, where he examined the idea that the faintness of the companion of Sirius was due to the fact that it had no intrinsic luminosity, but merely reflected the light of Sirius much as planets do in our solar system. This idea that Sirius B was a planet of Sirius was in vogue even before the companion was first observed. In 1855, Le Verrier spoke in terms of a planet when he discussed Bessel's unseen companion and it continued to be a credible explanation for the faintness of Sirius B throughout most of the 19th century. Observers would even occasionally refer to the companion as a "satellite" of Sirius. Gore was able to demonstrate that reflected light was not a reasonable explanation for the vast brightness difference, and that the companion had to be self-luminous.

In the section which he wrote on binary stars for *Astronomy*, Gore not only discusses Sirius A and B but goes well beyond the expected recounting of the then well-known properties of the system. He made several simple calculations of the relative brightness of the sun and Sirius A and concluded that stars of the "Sirian type" are intrinsically much brighter than the sun. Sirius B, on the other hand, must be much fainter than the sun in spite of the fact that it possessed a mass equal to that of the sun. In comparing Sirius A and B, Gore concluded, "The two bodies must, therefore, be differently constituted, and, indeed, the companion must be a nearly dark body." It was self-luminous, but only dimly so relative to Sirius. He therefore concluded it must shine with inherent light of its own, and it seems probable that it is a *large body*, cooling down and "approaching the complete extinction of its light." He then went on to speculate about the existence of planets around Sirius, but he was quite pessimistic that such planets could ever be directly observed. Clearly, Gore regarded Sirius B as a star of comparable size with Sirius A, but one having a much lower surface brightness. He never seems to have considered the possibility that it might actually be a much smaller star, and he did nothing further with his conclusions regarding the companion. Sadly, Gore died in his 76th year after being run down by a "jaunting" automobile in a Dublin street in 1910.

Gore's interpretation of Sirius B was very much in keeping with popular 19th century ideas about how stars evolved. The differences that were observed in intrinsic brightness and the color of the stars were imagined to be the consequence of a temporal evolutionary sequence. The most popular of these ideas was due to J. Norman Lockyer (1836–1920), an influential amateur scientist, who popularized a theory in 1890, which he called his "Meteoritic Hypothesis". His term "meteoritic" comes from his belief that meteorites crashing into the early sun supplied much of its initial heat. According to Lockyer, stars evolved from relatively cool red stars of immense size and shrank becoming hotter as they aged and the density of these contracting stars continually increased with time. At some point, when internal densities became too great for the perfect gas law to hold, the stars ceased contracting and cooled at a more or less constant radius and density. Thus, a given star would pass

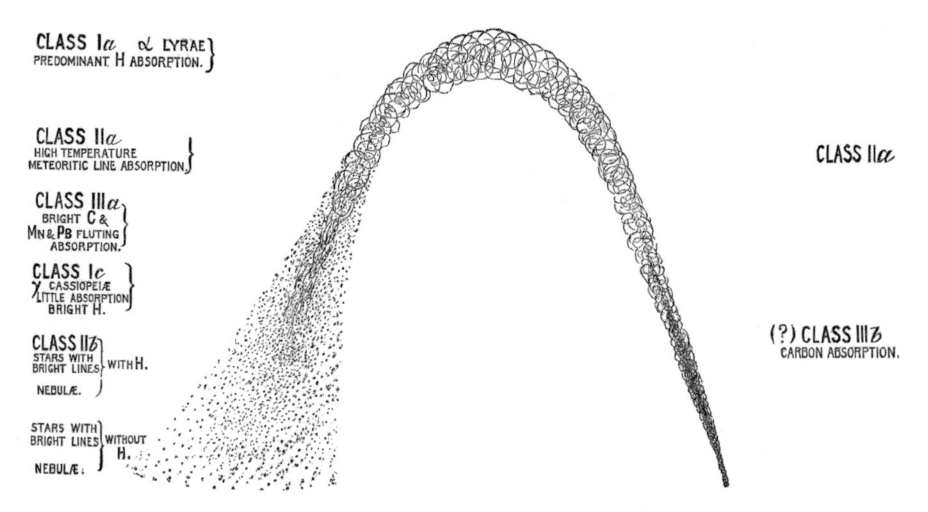

Figure 6.4. Norman Lockyer's concept of stellar evolution from 1890. The vertical annotation indicates how Lockyer envisioned stellar spectral classes changing with the thermal and temporal evolution of stars (the Meteoritic Hypothesis, 1890).

through a given temperature range twice as it aged (see Figure 6.4). First, as it was heating up as a large star and later as it cooled at a much smaller radius.

Sirius A and Sirius B seemed to fit this scenario perfectly. Sirius A as a hot blue star was near the peak of its brightness, while Sirius B had nearly completed the process and was nearing the extinction of its light. Even the relative masses of the two stars seemed to fit this theory. Sirius B, being the less massive star, cooled quickest while Sirius A, the more massive, had progressed more slowly and was now near its greatest luminosity.

A related idea was that the sun and the stars derived their energy from gravitational contraction. This concept was popularized by the German physicist Herman von Helmholtz in 1854 and was one of the first astronomical consequences of the discovery of the law of the conservation of energy. Helmholtz calculated that the sun's energy output could be accounted for by a slow gravitational shrinkage in its radius of 125 feet per year, an amount so small that it would not be observable. He could further estimate how long the sun had been shining at its present rate. He did this by calculating the total energy that it had derived over its lifetime from contracting from a very large radius to its current size, and dividing that total energy by its present rate of energy production. The answer that he got was 17 million years. Later physicists, such as Lord Kelvin in England, would refine the calculation to obtain an estimated age of 25 million years. Such young ages perplexed 19th century geologists, who required very much longer time scales for the geological processes on the earth, as well as early advocates of Darwinian biological evolution struggling to find adequate time for species to develop. Although it gradually became evident that the sun required a much larger source of internal energy than could be produced by gravitational attraction, it was not until the 1920s that nuclear energy came under serious consideration.

It was entirely possible to dispense with Lockyer's idea of meteorites crashing in to the sun, and adapt gravitational contraction as the sole source of solar energy, but retain Lockyer's scheme of stellar evolution. After all it was easy to show that if the sun derived its energy from the infall of meteorites then the subsequent increase in the sun's mass would produce changes to the orbits of the planets, which was not observed.

By 1890, nearly three decades after its discovery, Sirius B was rapidly approaching Sirius A. The elongated orbit would take the faint star to within just under three arc seconds of Sirius, where it would speed up and make a hairpin turn before heading back towards the point where it was first observed in 1862. The completion of a full orbit, or even the better part of an orbit, would give astronomers an opportunity to substantially improve the description of its path, and in particular its orbital period. As it approached Sirius, the companion was becoming increasingly harder to detect, let alone measure accurately. Consequently the number of reported observations began to fall dramatically. Sirius B was last seen on the evening of April 21, 1890, by the famous double star observer Sherwood Burnham, at the 36-inch telescope at Lick Observatory, at which point it had approached within 4.19 arc seconds of Sirius A. When Burnham tried to reobserve the star in the fall and winter of 1890, it could not be found. Burnham wrote, "I am surprised to find that the companion has passed beyond the reach of the large telescope. I had supposed that it would be measurable even at a minimum distance if the theory of its movement is substantially correct. It will probably not be seen for several years."

Although some of the basic elements of the orbit of Sirius B had been worked out by Safford, Auwers, and Peters even prior to its discovery, there remained much to be learned, particularly regarding the shape and orientation of the orbit. In order to establish these parameters the motion of two stars would have to be carefully observed over a substantial fraction of a complete orbit. Calculations of such orbits began to appear by the mid-1870s, but, as expected, they varied considerably. One significant prize that would result from a complete orbital solution (and an accurate parallax determination) would be an accurate determination of the masses of Sirius A and B, in terms of the sun.

The reemergence of Sirius B from the glare of Sirius was an event eagerly anticipated by astronomers world-wide in 1896. Its proximity to the various predicted orbital positions would provide a good indication of the accuracy of the numerous orbits which had been calculated in the decades since its discovery. Since the disappearance of Sirius B six years earlier, it was expected the companion would rapidly swing through a tight arc of nearly 180 degrees and reappear on the opposite side of Sirius A. Beginning early in the spring of 1896, large refractors all over the world were focused on Sirius in hopes of being the first to catch a glimpse of the reappearance of the faint companion. At the time, no sightings were reported before Sirius was lost in the glare of the sun in May. Sirius' reappearance in the morning sky in late summer and early fall of 1896 would be the next opportunity.

The first report of the reappearance of Sirius B came on September 12, 1896 from Thomas Jefferson Jackson See, at the 24-inch refractor at the Lowell Observatory in Flagstaff, Arizona. See (who we will later meet in connection with another curious aspect of Sirius), was an eager young astronomer and double star expert who would go on to hold a number of positions at various observatories during his long career and would, on more than one occasion, gain a reputation for being more ambitious than careful. See sent his findings to the *Astronomical Journal*, in a report dated October 6, 1896.

"Twenty minutes before sunrise on the morning of August 31, Mr. Douglass kindly interrupted his regular work on *Mars*, in order to enable us to make a search for the companion of *Sirius*. The air was perfectly steady and clear from recent rains, and the seeing was magnificent." See goes on to describe searching the field near Sirius for 5 minutes before inserting a diaphragm in order "to screen off the glare of the great star." "Within a few minutes I perceived a faint star of the 11th magnitude . . . ," See reported. He called his colleagues, including A. E. Douglas, who confirmed the reports and made measurements that were close to those of See. On three subsequent mornings, See and others, including Percival Lowell, observed the companion. Sirius B, it seems, was not in its predicted location. See estimated that he had observed Sirius B at a position angle of 220° (measured in a counter-clockwise angle from north) and 5.10 arc seconds from Sirius A. As he summarized: "It has fallen about thirty degrees behind its predicted place, and receded correspondingly further from the central star. This will necessitate a revision of the elements, which will have to have an increased period of about 54 years and a diminished eccentricity of approximately 0.5." See expressed his hope that he would be able to confirm his sighting when the observatory was moved (temporarily) to Mexico in a "couple of months".

Independent confirmation of See's observation was not forthcoming. In October of 1896, in the same volume of the *Astronomical Journal* containing See's report there appeared a one-sentence telegraphed announcement from Prof. Edward S. Holden, director of the Lick Observatory. It simply said, "Clark's companion to *Sirius* in its predicted position." This terse announcement served to call attention to a following account of observations of Sirius B by Robert J. Aitken and Professor Schaeberle which were much closer to predictions. Aitken stated "On Saturday morning, October 24th, I turned the large equatorial of this observatory upon *Sirius* and saw the companion." As Aitken dryly noted in his brief report transmitted on October 31, "Neither of us saw a star at the position given by Dr. See."

Aitken (1864–1951) provided further details of that morning, as he later recalled them in 1942. At age 31, Aitken had recently been hired at Lick at half the pay as an assistant astronomer and was given the opportunity to use some of the Lick telescopes for his own research, provided they were not in use by senior observers. After finishing a night's observing on the 12-inch refractor, Aitken passed by the floor of the 36-inch and stopped to inquire how the scheduled observer, Prof. Schaeberle, was "getting on". To his surprise the observation floor was empty, but the dome slit was open and pointed towards Sirius on the meridian. Not finding Schaeberle, Aitken required only a few moments to swing the telescope towards Sirius, where he was rewarded with a splendid view:

> "There stood the tiny companion as sharply and clearly defined as I have ever seen it, and both star images were perfectly steady. Without waiting for my notebook,

> I recorded a complete set of measures on the back of an envelope, determined my constants, and then turned up the telescope and went home happy in the thought that I, too, had seen the tiny companion on the first night of its emergence from the rays of its bright primary."

Aitken assumed that Schaeberle had been successful and simply retired early. As it happened the night had not gone well for Schaeberle, he had broken the micrometer's filament earlier in the evening, developed a violent headache, and finally retired after managing to reset a new spider web filament in the instrument; without making any measurement of Sirius. Aitken's casual measurements of that morning suddenly assumed great significance. Since Aitken was an unknown junior staff member, director Holden would not publicize observations without confirmation, which Schaeberle was able to provide the following night. Later, as he approached his 78th year, Aitken expressed the desire that he might, "live long enough to learn, after its emergence in about 1946, following pariastron passage, whether or not it will be found in its predicted place." He did survive to witness the appointed return of Sirius B.

Further confusion was added by Professor Stimson J. Brown of the U.S. Naval Observatory in Washington, D.C. After the reports from See and Aitken, Brown felt prompted to publish his description of a search for Sirius B he had conducted on the nights of March 13 and 14, 1896. Brown clearly regarded his observations as questionable and went so far as to directly quote comments from his logbook, "Seeing not very good; but at times image steady—at such times I saw persistently an unsteady, faint object just within the edge of the dancing rays ... Haven't much confidence until this result is verified." Brown's measurements were in better agreement with Aitken and Schaeberle than with those of See. Thus, Brown may well have been the first to detect Sirius B in March, but his observations were plagued by self-doubt. As additional observations accumulated in the spring and fall of 1897 it was soon clear that it was Aitken and Schaeberle who first reported the reemergence of Sirius B, that See's report was obviously spurious, and that Sirius B was close to its predicted position. The orbital period was now secure at 50 years.

The 19th century's final contribution to our understanding of Sirius came in 1898, with the first accurate measurement of its parallax. Sirius lies too far south for astronomers in most of Europe to have attempted to determine its parallax with good accuracy; however, it could be easily observed near the zenith from the Cape Observatory in South Africa. As mentioned in Chapter 5, Thomas Henderson had reported a parallax in 1839. However, the numerical value he quoted was 0.23 arc seconds, and his estimate of the uncertainty of the measurement was "not to exceed a quarter of a second". This was not improved upon until 1898 when another Scottish astronomer, David Gill, measured the parallax, with a heliometer (Appendix A) at the Cape Observatory, and obtained a value of 0.370 ± 0.011 arc seconds. Although Gill's value was only 50% larger than Henderson's, it had a dramatic effect on estimates of the total mass of the Sirius system, because from Kepler's third law the mass depends on the third power of the parallax. Bringing Sirius nearer by 50% has the effect of reducing mass estimates by over a factor of two. This is why early

19th century estimates of the mass of both Sirius A and B are often too large by this factor. A century later, the best available estimate of the parallax of Sirius, derived from a weighted mean between ground-based and spacecraft measurements, is 0.38002 ± 0.00128 arc seconds, in quite good agreement with Gill.

At the close of the 19th century both the orbit and the mass of the two stars were securely known. The true significance of the faint companion would, however, remain a riddle until the opening decades of the 20th century.

7

Giants among the Dwarfs

"What 'happens' is the stars"
—Sir Arthur Eddington, 1926

Today it is routine for astronomers to construct detailed theoretical models of the internal workings of stars and, with the help of modern computers, to calculate the past and present evolution of most stars. It is also possible to calculate a highly accurate theoretical description of how light emerges from a stellar atmosphere and how that light will subsequently appear when observed by telescopes at earth. To achieve this, all that is often required are some basic observational data, the brightness or distance of the star, its mass, its surface temperature, its surface gravity, its composition, all combined with a considerable amount of computation. These data are used as input to mathematical descriptions of the physics of a star and its atmosphere. It was not so long ago, however, that virtually all of this was unthinkable. In place of this present theoretical mastery of the stars there existed much confusion and a disheartening amount of basic ignorance. For example, it was only in 1939 that the key nuclear reactions which power stars were worked out, and sometime after that, that the pathways of stellar evolution were defined.

In the decades just before and just after the First World War there was a great flowering of observational and theoretical work that helped dispel many of the misconceptions that hampered our understanding of the stars. Much of the basic observational work was founded upon Harvard's massive *Henry Draper Catalog*, supplemented by observations made with the sequence of successively larger reflecting telescopes that George Hale was building in California. The theoretical work of interpreting these observations involved many people located at universities in Europe, the United States, and England. Three individuals stand out in these early efforts: a Danish chemical engineer with a passion for the stars, Ejnar Hertzsprung; a professor at Princeton University in New Jersey, Henry Norris Russell; and a

Cambridge don, Arthur Stanley Eddington. These three men played critical roles in elucidating the nature of stars and in particular in bringing to light the astonishing properties of these remote bodies. The critical observational and interpretative breakthroughs that changed our conventional view of the stars were primarily the product of the first two individuals, Hertzsprung and Russell, who worked independently and used a variety of different methods to gauge the distances to the stars. The final theoretical synthesis came from Eddington.

Good estimates of stellar distances are essential to the determination of the intrinsic luminosities of the stars. The luminosity of a star can be thought of as the total amount of energy that it radiates into space per unit time. A convenient unit of stellar luminosity is the total power radiated by the sun, which amounts to 2×10^{33} ergs s^{-1} (2×10^{27} watts). Likewise, a standard relative measure of stellar luminosity is the absolute magnitude of a star, or the magnitude it would appear to have, if the star were placed at a standard distance of 10 parsecs (33 light years). The absolute magnitude of the sun is 4.67; that is, it would appear this bright at a distance of 10 parsecs. Without a reliable means of estimating distance, two stars of the same apparent magnitude could lie at vastly different distances. For example, consider two stars with identical apparent magnitudes of 4.67 and spectral types similar to the sun. If one of the stars is just like our sun, then it would be located at a distance of 10 parsecs. If, on the other hand, the second star was a million times brighter (more luminous) than the sun, it would be located at a distance of 10,000 parsecs because the intensity of starlight diminishes with the square of the distance. The two stars could have similar spectral types in the *Henry Draper Catalog*, but would clearly be very different bodies. The more distant star would of necessity have a radius nearly a thousand times larger than the sun. That stars possess a considerable range of luminosities was already evident from the study of binary systems. Perhaps the best example of the extent of this luminosity range was Sirius. When John Ellard Gore in 1891 determined that the companion of Sirius was a nearly dark star, he also estimated the luminosity of Sirius itself to be 10 times that of the sun and the faint companion to be 1/500th that of the sun. Gore was off by a factor of two with respect to both Sirius and its companion, but at the time his result served as a concrete example of the luminosity differences that must exist among the stars.

By the beginning of the 20th century the distances to only a handful of stars had been determined from direct observations of their parallax. Indeed widely used textbooks such as Charles Young's 1891 edition of *General Astronomy* list the parallaxes of fewer than 30 stars, all measured by visual methods. Moreover, since the parallax methods of the time had a limited precision, they yielded meaningful results only for the nearer stars, such as Sirius. If progress in studying stellar luminosities was ever to be made, other indirect methods of estimating stellar distances beyond the effective limits possible with trigonometric parallax methods would be necessary. Fortunately, several indirect methods of estimating distances were being perfected at the time. Between 1889 and 1890 the Dutch astronomer, Cornelius Kapteyn (1851–1922) demonstrated that it was possible to deduce *average* distances for well-defined samples of stars from their *average* proper motions. The simple idea was that more distant stars had, on average, proportionally smaller proper motions. Another method was the study of nearby clusters of stars, such as the Pleiades and the Hyades. Such star clusters contain hundreds or thousands of stars that are grouped together in a small region of the sky. Members of such clusters also have similar proper motions; that is, they are moving through space as a physical group of related stars. In such clusters all stars have approximately the same distance from the sun and were formed at more or less the same time. Therefore, the study of such clusters simultaneously removes two troublesome and difficult variables, distance and age. The range of stellar luminosities in a cluster is directly related to the intrinsic luminosities of the stars. An additional powerful method is to intensively study binary systems, in particular eclipsing binary systems, where the stars are observed to mutually eclipse one another.

Ejnar Hertzsprung (1873–1969) took the first steps towards classifying stars on the basis of their luminosities. As a boy in Denmark, Hertzsprung developed a keen interest in astronomy, but following the wishes of his father he studied the more practical and lucrative subject of chemistry. After graduating from the University of Copenhagen in 1898, he became a chemical engineer and briefly worked in Russia and Germany, before returning to Copenhagen in 1901. It was back in Copenhagen that Hertzsprung began to make use of the telescopes at the University and Urania Observatories in Copenhagen. During this period, as effectively an "amateur astronomer", Hertzsprung published several key papers on stellar luminosity. However, his insights went largely unnoticed due to his lack of a professional position, and to the fact that the papers were published in an obscure German journal dealing with photography. Hertzsprung did, however, attract the attention of the gifted and versatile German astronomer Karl Schwarzschild (1873–1916) and in 1908 Schwarzschild invited him to work at Göttingen in Germany. After only a few months, Hertzsprung was appointed to a professorship in Göttingen, and later in 1909 Hertzsprung followed Schwarzschild to Potsdam in Germany. Hertzsprung worked throughout the First World War at Potsdam, and in 1919 accepted a lifelong position at the Leiden University Observatory in Holland.

In 1905 Hertzsprung (Figure 7.1) had published a brief paper in which he first called attention to the fact that the stars with narrow, sharp spectral lines—those which Antonia Maury had assigned to her subclass "c"—all seemed to have proper motions that were almost too small to measure. Hertzsprung concluded that these stars must lie at vastly greater distances than those of the more numerous "a" and "b" subclasses, which had relatively large proper motions. Since these more distant "c" subclass stars also had relatively bright apparent magnitudes, they must also be incredibly luminous. In effect, Maury's "c" subclass stars were indeed unique objects after all. Hertzsprung followed this with another paper in 1907 strengthening his argument that the "c" subclass stars were fundamentally different from other stars. Because these stars could be seen over great distances, they had to be very sparsely distributed in space, compared with normal stars like the sun. He also observed that "bright red stars"—like Arcturus, Regulus, and Betelgeuse—were rare per unit volume of space and consequently represent a brief, short-lived stage of stellar evolution.

Figure 7.1. Ejnar Hertzsprung (Yerkes Observatory photograph).

Another method used by Hertzsprung to study stellar luminosities was that of observing stellar clusters. In 1911 he published diagrams showing the relationship between the color and the luminosity of the stars in the Pleiades and Hyades star clusters. The measures of "color" used by Hertzsprung were derived from the relative brightness of the stars as photographed through blue and yellow filters and also from the effective wavelengths at which the stars appeared to be brightest in objective grating spectra. Both of these measures are basically proxies for stellar temperature. The diagrams clearly demonstrated that most stars fell along a relatively narrow band which he called the "main sequence". Quite apart from the main sequence were a small number of very luminous red stars.

Two years earlier, in 1909, Hertzsprung had proposed that Sirius was a member of the Ursa Major moving star cluster. It had earlier been noted that five bright stars in the constellation of Ursa Major (the Big Bear) had nearly identical proper motions. These stars form the lower part of the handle and the ladle of the familiar "Big Dipper". If the proper motions of the stars in a moving cluster are projected backwards in time, their paths converge and appear to originate from a single point in the sky. Thus, they must have had a common origin in the distant past. Hertzsprung noted that Sirius seemed to share the proper motion of the Ursa Major stars. Nominating Sirius for membership in the Ursa Major moving cluster, however, represented a tremendous extrapolation. Sirius is securely located in the southern hemisphere, while

Figure 7.2. Henry Norris Russell (Yerkes Observatory photograph).

Ursa Major is located high in the northern sky, over 60 degrees away. Nevertheless, when Hertzsprung tracked back the motion of Sirius, he found that it passed within 1/10th of a degree from the projected point of origin for the Ursa Major moving cluster, located in the constellation Lynx. As will be discussed in Chapter 12, modern estimates for the age of Sirius indicate that it may not be the same as those of the Ursa Major stars, and thus Hertzsprung's proposed link to the Ursa Major moving cluster may be merely coincidental.

Hertzsprung's main competitor in the effort to understand the luminosity of the stars was Henry Norris Russell (1877–1957). Russell (Figure 7.2) was born into a well-to-do family on Long Island, New York. Henry, the eldest of three boys, was a serious bookish lad, imbued by his father, a Presbyterian minister, with a compulsive work ethic and strong religious devotion which he maintained throughout his life. From his well-educated mother, Eliza Norris, Henry acquired a love of learning and clear knack for mathematics. At the age of 16 he entered Princeton University, where he followed the standard liberal arts curriculum of the day with its strong emphasis on

Greek, Latin, and the classics. It was mathematics and physics at which he excelled, however.

Russell's first formal encounter with astronomy came during his junior year at Princeton with the introductory astronomy course taught by Charles A. Young (1834–1908). Young was an inspiring teacher and enthusiastic proponent of astronomy. His textbook *Manual of Astronomy* remained in wide use at American colleges and universities for over thirty years during the period before the First World War. The astronomy practiced by Young was not the positional and mathematical variety favored by many American astronomers of the day, but rather focused on the newer field of spectroscopy and its application to the physical properties of the sun, the stars, and the planets: in short he was both an advocate and practitioner of the up and coming field of astrophysics. Russell graduated in 1897, at the age of 20, with the highest honors. His Princeton yearbook photograph shows a very earnest bespectacled young man with his hair parted right down the middle. It is hard to imagine Henry ever having much contact with the boisterous social life of turn of the 19th century Princeton undergraduates.

Russell spent the years of 1902 to 1905 at Cambridge University in England studying the theories of George Darwin (Charles Darwin's son) having to do with the stability and centrifugal breakup of rotating fluid bodies. Darwin's theories were popular at the turn of the last century and were invoked to explain the origin of the earth–moon system and the formation of close binary stars. At Cambridge, Russell also worked intensively on the practical problem of developing methods to photographically determine stellar parallaxes. Both of these interests were to significantly influence the course of his early research. Following a long recuperation from typhus, he returned to Princeton to take up a junior faculty position in the fall of 1905. After the retirement of Young, Russell became part of a two-man astronomy department, with modest teaching responsibilities that left him time to measure parallaxes from the photographic plates that he had obtained in Cambridge. Over the next several decades Russell became one of the leading astronomers in the United States. He would spend the remaining five decades of his life, dividing his time between Princeton and the family estate on Oyster Bay, Long Island, New York.

One of the great milestones leading to an understanding of the stars grew out of the work of Russell and Hertzsprung. Independently, using different methods, they established the fundamental relationships between the luminosity of the stars and their spectral types. Russell was a superb analyst and synthesizer of astronomical data and an omnivorous consumer of stellar distances. Initially he made use of the stellar parallaxes that he and others had measured. When these proved insufficient he turned to other indirect methods of distance estimation. Chief among these was the use of eclipsing binary stars, whose orbits were precisely inclined so that when the stars passed in front of, or eclipsed one another, the total light would first dim and then recover. It was from eclipsing systems that Russell was first alerted to the fact that some stars must be immense and possess extremely thin distended atmospheres. In order to extract information from such systems Russell pioneered effective methods of analyzing the eclipse light curves to determine the masses, radii, and luminosities of the stars. Russell also applied many of the same statistical techniques used by

Hertzsprung to estimate stellar distances. Unlike Hertzsprung, however, much of Russell's work focused primarily on determining the luminosity of individual stars rather than on statistical averages.

Distances were not all that Russell needed. He also required spectral types for his stars and thus he forged an early relationship with Edward Pickering to gain access to the preliminary results from the monumental *Henry Draper Catalog* (completed in 1924) that was producing spectra for virtually all stars brighter than 9th magnitude. In the period between 1909 and 1913 Russell began to assemble his luminosities and to explore the relations between luminosity and spectral type.

When Russell plotted the luminosity, derived from his trigonometric distances, against the Harvard spectral types (Figure 7.3), he noticed that the stars naturally grouped themselves into two distinct populations. One group, containing most of the stars, clearly ran in a diagonally downward sloping band, from left to right, across his plot. These were stars like the sun whose luminosities decreased steadily as the spectral types progressed from B to M, along the Harvard spectral sequence. The second was a less distinct group consisting of a sparse sprinkling of stars running horizontally along the top of the plot, well above the first group. These were stars, like Betelgeuse, that were much more luminous and distant. The two populations seemed to merge for the most luminous blue stars near spectral types B and A.

From his studies of binary stars Russell knew that these two types of stars must have also very different mean densities. The solar-type stars had high average densities, while the luminous stars included stars he knew exhibited very low density. Initially Russell interpreted his plot as a dramatic confirmation of Norman Lockyer's picture of stellar thermal evolution, where stars of initially large size contracted and heated and then, reaching a peak temperature, cooled to smaller denser stars as they radiated away their finite amount of gravitational energy. Physically, it seemed to make sense, the large stars or "giants" as they became known, were very distant and had to be larger low-density stars in order to be as bright as they appeared. Nearer fainter stars, or "dwarfs" as they became known, of the same spectral type must have similar temperatures but were much fainter and therefore smaller and denser. The dichotomy reached its most extreme for the red M-type stars such as Betelgeuse which are hundreds of times larger than the sun, and some nearby faint red M-type stars which had to be much smaller than the sun. The manifest difference between the two stellar types is evident in the fact that many remote red giant stars can easily be seen with the naked eye in the night sky, while it takes a telescope and a star map to view any of the vastly greater number of faint red dwarfs that inhabit the solar neighborhood.

During the summer of 1913 Russell presented various versions of his data at several conferences. In London, at a meeting of the Royal Astronomical Society, he gave a talk on the relation between stellar luminosities and spectral types. The title of the paper, as it appeared later in the *The Observatory*, was *"Dwarf" and "Giant" Stars*. Russell first used these terms in the talk, but mistakenly attributed their origin to Hertzsprung, who had only used the German word "Giganten". The phrase "giants and dwarfs" quickly entered the vocabulary of astronomers. It was suddenly evident that not only did stars follow a spectral sequence, but when luminosity was

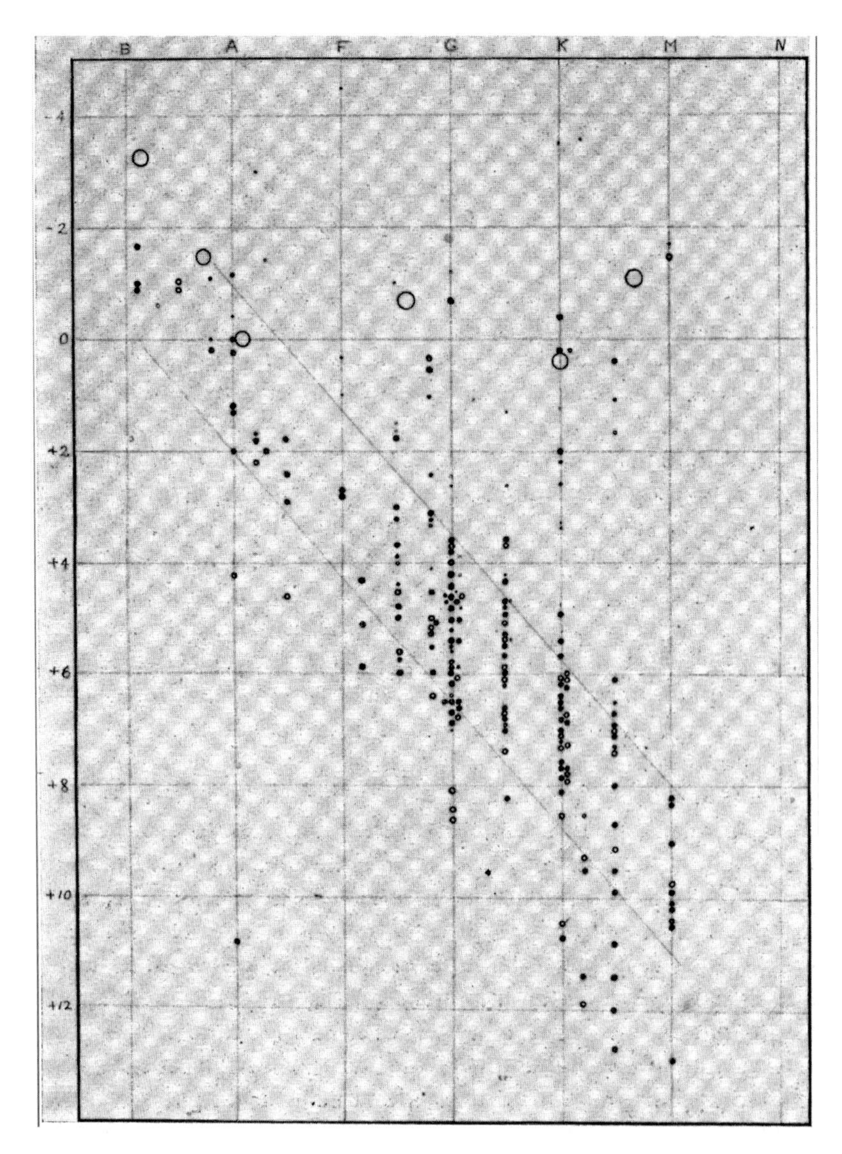

Figure 7.3. Henry Norris Russell's 1913 plot of spectral type vs. luminosity. The white dwarf 40 Eridani B is the isolated point in the lower left-hand corner of the plot (*Popular Astronomy*, Vol. 22, 1914).

considered, the stars also fell into two different classes. This classification was commonly referred to as the "theory of giants and dwarfs". Russell presented the classic form of his diagram (shown in Figure 7.3) at the December 1913 meeting of the American Astronomical Association in Atlanta, Georgia.

That such vastly different kinds of stars seemed to exist was now clear from the work of Hertzsprung and Russell, and independent confirmation was not long in coming. In 1914, Walter S. Adams (1876–1956) and Arnold Kohlschütter (1883–1969), working jointly at the Mt. Wilson Observatory, published a paper in which laboratory results on the intensity ratios between certain spectral lines were directly linked to the observed intensity ratios of these lines in stellar spectra. The spectra of giant and dwarf stars showed marked differences in these line ratios which could be explained if the lines were formed in stellar atmospheres at low pressures (giant stars), or at higher pressures (dwarf stars). This was an additional observable distinction between the two types of stars and became a useful technique to estimate stellar luminosities called the "spectroscopic parallax".

As far as the public was concerned, the most dramatic confirmation of the existence of giant stars came in 1920 when the American physicist, and Nobel Prize winner, Albert Michelson (1852–1931) used the Mt. Wilson 100-inch Hooker telescope to actually measure the diameter of Betelgeuse with his ingenious stellar interferometer. Michelson completely covered the primary mirror, save for two holes near its periphery. He then attached a 20-meter steel beam to the Hooker telescope, which projected outward from the front end of the telescope structure. At the ends of the beam he placed two flat mirrors and directed their light paths to two similar flat mirrors midway along the beam, so that light was directed downwards towards the holes in the primary mirror cover. Effectively, Michelson used the large telescope mirror to focus the separate beams of light from the two outrigger mirrors onto a specially constructed eyepiece. When the arrangement was in nearly perfect alignment he could view a blurry patch of light from a star, on which were superimposed faint bands. These bands were interference fringes produced by the two separate light beams and were formed when the lengths of the two light paths differed by precisely an integer number of wavelengths. By adjusting the length of the light path Michelson could make the fringes appear and disappear. The relationship between the wavelengths of the light and the geometry of the light paths yielded the angular diameter of the star. Michelson and his colleague, Pease, measured the angular diameter of Betelgeuse to be 0.047 arc seconds and calculated its diameter to be over 550 times the diameter of the sun. The diameters of several other giant stars were measured with this instrument; however, it proved very difficult to control and the technique was abandoned.

The discovery of giant stars eventually led to a significant revision in Annie Cannon's Harvard classification system, described in Chapter 6. The basic letter and number scheme, which essentially follows a temperature sequence, was retained, but to it was added a Roman numeral to indicate to which luminosity class a star belonged. Dwarf stars like the sun and Sirius were assigned a "V" so that they became G2V and A0V, respectively. The giant stars were expanded into a set of subclasses IV, III, II, and I, in order of increasing luminosity. The stars of subclass IV are called subgiants; subclasses III and II became the giants. Subclass I was reserved for stars that became known as supergiants. For example, such familiar stars as Betelgeuse and Arcturus are now designated M2I and K2III, respectively, so that Betelgeuse is designated a supergiant and Arcturus a giant star. We now know that each of the luminosity classes represents a different stage of stellar evolution and that these stages occupy different regions of what has become known as the Hertzsprung–Russell, or

H-R diagram. This joint designation, which came into being in the 1930s, recognizes the contribution that both Hertzsprung and Russell made to our understanding of the stars.

Today the H-R diagram not only depicts the static distribution of stars in temperature and luminosity but also serves as a map of the evolutionary relationships between stars of different luminosities and temperatures. For example, the dwarf stars, which Hertzsprung first called the main sequence stars, are now known to represent stars of different masses which are in the hydrogen-burning phase of their evolution. Such relationships, however, were not understood when Hertzsprung and Russell first developed their ideas. The H-R diagram has also become an all purpose tool for stellar astronomers; consequently it has many manifestations. For example, temperature, spectral type, and color can all be plotted against various representations of luminosity such as absolute magnitude or the total luminosity for a star.

The initial interpretations of the H-R diagram followed 19th century ideas of stellar evolution. The most popular of these, and the one initially favored by Russell, was due to the ideas of two people. In 1869 a former U.S. Patent Office employee, Homer J. Lane (1818–1890), published a key paper in which he investigated the distribution of temperature, pressure, and density in the solar interior. He began by specifying that the sun was in gravitational equilibrium, with the internal gas pressure balancing the gravitational pressure at each point within the sun. He also assumed that the matter inside the sun obeyed the ideal gas laws, where the gas pressure is proportional to the temperature and density. The sun's energy, he believed, was conveyed to the surface by the convection of hot gases. He expressed all of this in a differential equation which he numerically solved in order to estimate the temperature of the solar surface. The answer he got was a surface temperature about six times what we know now to be the correct answer. Nevertheless, his basic ideas were employed well into the beginning decades of the 20th century to build interior models of the sun and the stars, which were called "polytropes" (Appendix A). Most astronomers, however, viewed these stellar models as theoretical exercises, which had little application to real stars, since it was believed that the density of matter in stellar interiors was simply too high for the ideal gas law to hold.

The popular ideas about how stars evolved were less mathematical and more qualitative. Normal Lockyer's "Meteoritic Hypothesis" (Chapter 7) provided for many a convenient pictorial representation of how stars evolve and also why stars of the same spectral type and temperature can come in two different sizes. As mentioned, according to Lockyer, the evolution of stars proceeded from large cool stars that became hotter and brighter as they aged and shrank. Stars would reach a maximum luminosity near the point where contraction diminishes due to increased density. A long period of cooling then ensues at a more or less constant radius and density. Thus, each star passes through a given temperature range twice as it ages. It is easy to see how Russell and others could interpret the existence of giant stars and dwarfs according to Lockyer's theory. By 1920, however, there was one vocal critic who dissented vigorously and eloquently from the idea that the stars drew their energy from gravitational contraction and from the view that stellar evolution proceeded in the manner depicted by Lockyer.

Arthur Stanly Eddington (1882–1944) was born into a well-to-do Quaker family in the English Lake District. At the age of two his father, a local headmaster, died and Arthur went with his mother and sister to live with relatives in Somerset in southwest England. By all accounts Arthur was a precocious child who was a careful observer of nature and a compulsive counter of things, such as the number of words in the Bible and the stars in the sky. He excelled at secondary school and entered a college in Manchester at age 16, where he graduated four years later with top honors in nearly all subjects. Eddington earned a scholarship to Trinity College at Cambridge in 1902, where in his second year he achieved the top rank of Senior Wrangler in mathematics, the first person ever to gain this distinction at such an early age. Eddington would spend nearly all of his remaining forty-four years at Trinity, first as a student, then as a fellow, and finally as the prestigious Plumian professor of Astronomy and Director of the Cambridge Observatory.

During his first two decades at Cambridge, Eddington would solve one of the most vexing problems of astronomy in the early 20th century: What goes on inside a star? He was also an early and compelling proponent of Albert Einstein's theory of General Relativity and wrote one of the most lucid accounts of the theory available at the time. In 1919 he became a public figure when he led the British solar eclipse expedition that verified Einstein's prediction of the bending of the path of starlight by the gravitational field of the sun. Eddington was primarily a theoretical astronomer, who turned abstract ideas into concrete mathematical models of how stars and other bodies behave. Nevertheless, he was also acutely aware and intimately familiar with the new discoveries and measurements in atomic and subatomic physics that were emerging from the Cavendish Laboratories on the Cambridge campus and elsewhere in Europe and America. He was one of the first to apply these discoveries and data to problems in astronomy. By the mid-1920s he had become an international scientific celebrity and statesman, who found time to write popular accounts of the exciting new discoveries being made in astronomy. He could charm lay audiences with his wit, and masterful use of descriptive language could overwhelm his scientific colleagues with his authority. There is perhaps no more quotable astronomer in the English language than Eddington. Ironically, at times his class lectures could be as dull as dishwater and he could be absent-minded and self-absorbed to the point of rudeness. As will become evident, he was also supremely self-confident and annoyingly stubborn, if he believed he was correct. During his lifetime he accumulated nearly every available award, with the exception of the Nobel Prize in Physics, for which his work on stellar interiors clearly qualified him.

Arthur Eddington was among those in the audience at the Royal Astronomical Society in 1913 when Russell presented his data on the existence of giant stars. Soon thereafter, it occurred to Eddington and others that if there were stars whose densities were thousands of times lower than that of the sun, then perhaps the ideas of Homer Lane and the subsequent polytrope models could be relevant to these stars. After all, there was no question that the ideal gas law would apply at such low densities.

The sun is centrally compressed to an astonishing degree. Its mean density, the total mass divided by the total volume, is only 1.4 $\mathrm{gm\,cm^{-3}}$, about the same as coal or rock salt. However, this does little to convey how mass is actually internally distributed within the sun. The visible photospheric layers have a density less than that of air, while in the center the density is 160 gm cm^{-3}, more than five times the densest materials on earth. A density of 1 gm cm^{-3} (that of water) is only reached midway between the surface and the center, and the central 30% of the sun contains 2/3 of all the mass. This region produces virtually all the sun's energy. The fact that the densities in the sun manage to reach very high levels was a major astronomical dilemma at the beginning of the 20th century, following the discovery of the atomic nucleus surrounded by electrons. The theory developed by Lane in 1869 assumed that the interior of the sun behaved like a perfect gas. Yet physicists and astronomers knew that gases ceased obeying the perfect gas law long before they reached such high densities. The material inside the sun must be compressed to such an extent that the individual atoms were in close contact with one another and therefore they must behave like a solid or, given the high temperatures, perhaps some form of super-dense, hot fluid. Energy from the center of the sun was then assumed to flow outward via the convective motions of this fluid.

Eddington, a Quaker and a conscientious objector, set out during the dark days of the First World War to build a credible mathematical model of the interior of a giant star. In the process he intensively investigated the processes that must be important within a star. One long-overlooked consideration was the importance of radiation to the structure of a star. Gas pressure had been long considered to be the dominant factor in the structure of a star, and a star's internal heat was thought to reach its surface primarily as mentioned through convection. Eddington showed that a star's internal energy flowed outward primarily through radiation. Moreover, the flow of radiation was so strong that it exerted a pressure that helped to support a star against gravity. It had long been known that light was capable of exerting a feeble pressure against matter that was proportional to its intensity. It was Eddington, however, who first realized that the intensity of radiation (in the X-ray band) inside a star could become so high that the resulting pressures would rival or even overwhelm the internal gas pressures. Eddington likened this outward flow of radiation to a fierce central wind blowing from within a star. Gas pressure and radiation pressure behave differently inside a star. For the gas component, the pressure (P_g) is proportional to the temperature (T), ($P_g \propto T$) while the radiation pressure (P_r) is proportional to the fourth power of the temperature ($P_r \propto T^4$). Therefore, at low temperatures gas pressure dominates, while at high temperatures radiation pressure dominates. It is in the intermediate regime, where both gas and radiation make significant contributions to the total internal pressure, that Eddington found most stars seemed to exist. In fact he concluded that this regime of temperature and pressure explained why stars exist at all and why they were as large and massive as they were. He required, however, a theory to calculate how radiation and matter interacted to produce the necessary pressure.

Eddington was aided in his study of stellar interiors by developments in early 20th century physics. In particular, he adopted new ideas of the structure of the atom, most notably by Bohr's theory of electrons occupying various energy levels in an atom. A critical milestone was the theory of atomic ionization developed in 1920 by a young Bengali physicist, Merghnad Saha (1893–1956), working in India. Saha employed the mathematical language that the physicists James Maxwell and Ludwig Boltzmann had used some fifty years earlier to describe the thermal equilibrium of atoms and molecules, and extended it to include a description of the ionization equilibrium of atoms and electrons in a hot ionized gas, or plasma. At a given temperature and pressure, each particular chemical element would, on average, retain a certain fraction of its electrons and contribute the rest to a "gas" of free electrons. Eddington also made use of the new atomic theory and laboratory results on the absorption of X-rays by matter. When he applied these ideas, and the corresponding data, to the interior of a star he found that he could calculate the "resistance" experienced by radiation as it worked its way from the interior to the exterior of a star. This "resistance" is called opacity and Eddington found that, over a wide range of conditions, it could be characterized by a relatively simple mathematical expression.

In the end Eddington was able to express the luminosity of a star in terms of its mass and the mean molecular weight of stellar matter. This represented an astonishing simplification of the apparent complexity of a stellar interior. The luminosity still depended, however, on an unknown numerical factor involving the opacity of stellar matter and the only way of determining that factor was to apply the equation to a real giant star, of known mass and luminosity.

When Eddington finally completed a self-consistent analysis of the interior of a giant star, he had very few real stars with which to compare his theory. Most giant stars lay at great distance and their masses, radii, and luminosities were poorly known. Eddington was fortunate. The star Capella, the brightest star in the constellation Auriga, is actually a double star system containing two giant stars with temperatures similar to the sun. The two component stars, however, had never been resolved. In 1921 the American astronomer J. A. Anderson, using a modification of Michelson and Pease's stellar interferometer on the 100-inch telescope at Mt. Wilson, made a series of six observations of Capella during its 104-day orbital period and succeeded in resolving the two stars at a separation of 0.0536 arc seconds and determining the orbit of the system from which it was possible to measure the masses of the two stars, which are 4.18 and 3.32 times that of the sun. Eddington had, in Capella's brightest and most massive component, just the star he needed to calibrate his theory.

When he applied the theory to other giant stars he found a very satisfying agreement between theory and observation. This gave him confidence he was on the right track, at least for the giant stars. In February of 1924, as he was preparing to publish his results, he decided to investigate how much the denser dwarf stars actually deviated from his theory. There existed a great deal of good observational data on the dwarf stars, including the sun, whose mass, radius, and luminosity were known very precisely. Eddington expected to see an increasing deviation between the predicted and observed luminosity of the dwarf stars, as their densities increased. To his utter astonishment, his theory fit the dwarf stars even better than the giants!

It didn't take Eddington long to determine what was happening. At the high temperatures and densities inside the dwarf stars the individual atoms were virtually stripped of their outer electrons and retained only the innermost electrons in tight orbits near the nucleus. Effectively, these strongly ionized atoms were now much "smaller", and able to pack themselves much closer together than their neutral counterparts. Densities could be much higher and yet the matter could still behave as an ideal gas.

Eddington summed up all of this in a famous 1924 paper entitled *On the Relation between Masses and Luminosities of the Star*. In this paper Arthur Eddington first set forth the fundamentally important relation between the mass of a star and its luminosity. Thus, it is a star's mass that is the chief factor in determining its luminosity and spectral type. It had been earlier noticed that when a pair of binary stars had similar masses, both tended to have similar luminosities and that both the luminosities and masses of most red stars declined with increasing redness. However, it was Eddington who explained why this occurred by demonstrating the conditions under which stable stars can exist, as a function of mass.

The realization that it was a star's mass that determined its luminosity was a milestone in understanding the nature of stars. It was also very much contrary to the prevailing views of the late 19th and early 20th centuries, which held that a star's luminosity varied significantly over its lifetime. For example, a star such as Sirius would progress from a large cool red star and, through contraction, heat up, and then, near its maximum luminosity, glow bluish white. At the point when its density could no longer increase it would begin to cool. Eddington, by showing that the matter in a star continued to behave as a gas, even at very high densities, effectively refuted this earlier theory of stellar evolution. In its place, he showed that most of his stars remained relatively stable with constant luminosities, radii, and temperatures, all of which was determined by a star's mass, a factor which could hardly change by much over the life of a star. Thus, Eddington's stars were effectively static, and without knowing the mechanism of how energy was produced in the interior of a star he could only guess at how the stars might evolve.

In effect Eddington was the first to peer into one of nature's most secret and hidden domains, the interior of a star. In his own words what existed there was a "hurly-burly" of atoms, electrons, and photons in constant collision—electrons were being rapidly stripped from atoms and just as rapidly reattaching themselves to new atoms in a maelstrom of violent collisions. The atoms and electrons were bathed in a sea of photons which were continuously being emitted and absorbed. The ultimate source of all this energy was not precisely known at the time, but Eddington suspected as early as 1920 that the source was nuclear reactions or "subatomic energy", as he called it. However, at the time no one could describe exactly how this might be done. He was one of the first to suspect that stars were powered by the conversion of hydrogen into helium and even suggested that nuclear reactions in the stars were the source of all elements heavier than hydrogen. In the end, Eddington did not need to know how the energy was produced, as long as the energy source was constant and confined to a very small region of the stellar interior, or core. It was enough to simply require that the energy produced per gram per second by material inside a star matched the energy radiated away at the surface. The energy output it turned out was controlled by the mass of the star and its subsequent gravitational pressure. This was just one of the grand simplifications that allowed Eddington to proceed.

Eddington summarized most of his work on stellar interiors in a classic 1926 book with the authoritative title *The Internal Constitution of the Stars* (a title that he recycled from a remarkably prophetic address he first gave at a meeting of the British Association for the Advancement of Science in 1920). This book swept away most of the old misconceptions about the stars, how they were related to one another, and how they were internally constructed. Although the book contains much detailed mathematics, a great deal of it, particularly the early chapters, are quite approachable even to lay readers. It quickly became a 20th century scientific landmark and helped establish Eddington as one of the preeminent scientists of his era. So confident was Eddington in his new theory of the stars that in the book he imagined the existence of a hypothetical physicist on a cloud-shrouded planet, unable to view the heavens. Using his new ideas Eddington showed how the physicist could, without ever seeing a real star, deduce their existence! Eddington began by considering balls of gas with different masses, starting with an insignificant ball having a mass of 10 grams. The mass of each successive ball of gas increased by a factor of 10, all the way up to 10^{40} grams (the sun, for example, has a mass of 2×10^{33} grams). The gas balls are confined by their own self-gravity and the physicist calculates the relative contributions of the internal gas pressure and the radiation pressure. At smaller masses, gas pressure totally dominates and the putative "stars" slowly contract, radiating away their gravitational energy, never becoming real stars. At around 1/10 the mass of the sun, however, gas pressure still dominates, but radiation pressure starts to become important, and a stable star can exist. When the masses become too large, at around 100 solar masses, radiation pressure totally dominates, the stars become unstable and literally blow themselves apart, not necessarily violently but they expel their outer layers in a fierce stellar wind (Eddington imagined that they might become rotationally unstable and break up into less massive stars). Stable stars, Eddington (and the hypothetical physicist) concluded, can only exist in a narrow range of masses between about 1/10th and up to about 100 times the mass of the sun. This is just the range of masses that astronomers such as Russell were finding for real stars. The current best estimates range from about 7% the mass of the sun up to about 120 times the mass of the sun, not far from Eddington's initial estimates. At the end of his parable of the hypothetical physicist, Eddington imagined that the clouds cleared and the physicist could look up and view real stars for the first time.

With Eddington's masterful explanation of how nature managed to construct dwarf and giant stars, the central astrophysical problem posed by the variety of stellar spectra and luminosities seemed to be largely solved. It all had to do with mass. Exactly how stars produced their energy and how they evolved was not known, but Eddington had ignominiously buried the old idea that stars evolved by shrinking and becoming denser. Yet, hidden within the stellar luminosity data of Hertzsprung and Russell, lurked yet a third type of star, one with even more unimaginable properties.

Beginning in 1910 the real nature of Sirius' faint but massive companion slowly began making itself apparent to the 20th century astronomical community in a process that unfolded over the next two decades. The first inkling that there might be something truly peculiar about Sirius B actually began with an unexpected discovery involving another faint star in the constellation Eridanus (The River), 40 Eridani B. Like Sirius B, 40 Eridani B is a faint member of a star system, but unlike Sirius B, it is not deeply enshrouded in the glare of its primary, so that observations of both stars are possible without great difficulty. The primary star 40 Eridani A (or σ^2 Eridani, as it is also designated) is a rather unremarkable 5th magnitude K star, somewhat cooler and less luminous than the sun. Its 9th magnitude common proper motion companion (40 Eridani B) had been known since its discovery by William Herschel in 1783. (Herschel observed a close pair of stars 40 Eridani B and a fainter red star, 40 Eridani C. Together this pair forms a triple system with 40 Eridani.) The primary star, 40 Eridani, had a well-determined parallax, which implied a distance of about 15 light years and a luminosity that fit well with values that Russell was obtaining for similar dwarf stars. The critical discovery that first prompted Russell's curiosity, and later spurred his doubts, about the companion, 40 Eridani B, occurred in 1910. He recounted several versions of this event during the early 1940s; the following account was published in 1944, while Russell was working for the Army Air Corps during the Second World War:

> "The first person who knew of the existence of white dwarfs was Mrs. Fleming; the next two, an hour or so later, Professor E. C. Pickering and I. With characteristic generosity, Pickering had volunteered to have spectra of the stars which I had observed for parallax looked up on the Harvard plates. All those of faint absolute magnitude turned out to be of class G or later. Moved with curiosity I asked him about the companion of 40 Eridani. Characteristically, again he telephoned to Mrs. Fleming who reported within an hour or so, that it was of Class A. I saw enough of the physical implications of this to be puzzled, and expressed some concern. Pickering smiled and said, 'It is just such discrepancies which lead to the increase of our knowledge.' Never was the soundness of his judgment better illustrated."

The Mrs. Fleming Russell mentioned was Williamina Fleming who had developed the initial Harvard spectral classification system (Chapter 6). Both 40 Eridani A and 40 Eridani B are on the same Harvard photographic plate, respectively designated as HD 26965 and HD 26976 in the *Henry Draper Catalog*. What puzzled Russell and Pickering at the time was why a star with spectral "class A", clearly at the same distance as the primary 40 Eridani A, should appear so much fainter than the primary of spectral "class G" (the modern classification is K1V).

Three years later, in the spring of 1913, Russell prepared the first version of what would become his signature plot of stellar luminosity vs. spectral type. In that plot (see Figure 7.1), some 11 magnitudes below the other stars of "type A", sat a lonely but conspicuous point, 40 Eridani B. Russell's attitude at the time was to disown the star by doubting the spectral classification. In June of 1913 at his talk in London to the Royal Society he spoke of the spectrum as being "very doubtful". Later in December 1913 at the Atlanta meeting of the American Astronomical Society he commented "the single apparent exception is the faint double companion to σ^2 Eridani, concerning whose parallax and brightness there can be no doubt, but whose spectrum, though apparently of Class A, is rendered a very difficult observation by the proximity of its far brighter primary." Russell was being deliberately coy, there was no serious problem due to the brighter primary; he simply doubted the reality of Fleming's classification of the star.

Shortly after Russell's talk in London he traveled to the continent for more scientific meetings. In Hamburg he met for the first time with Ejnar Hertzsprung and showed him a preliminary version of his plot of stellar luminosities. The two men also discussed the dilemma of 40 Eridani B. Hertzsprung was as perplexed by the luminosity and classification of 40 Eridani B as Russell. Hertzsprung maintained that A stars were invariably luminous objects and Russell had categorically stated "There do not seem to be any faint white stars."

Confirmation of Fleming's classification arrived just a year later when Walter Adams at the Mt. Wilson Observatory succeeded in getting a spectrum of 40 Eridani B with the 60-inch telescope in the fall of 1914. Adams published a very brief note that October making clear there could be no question that, "The spectrum of the star is A0." When Hertzsprung read of this result in Potsdam he wrote to Adams on November 14, commenting that he found that "this absolute dark A-star so far off the rule is very interesting." In the letter Hertzsprung also noted that he had independently obtained evidence concerning the color of 40 Eridani B. In 1912 Hertzsprung had spent five months at Mt. Wilson Observatory, obtaining "effective wave-lengths" of a number of stars with well-known parallaxes. His technique involved the use of a coarse transparent grating covering the telescope aperture that produced small right- and left-hand dispersed images on a photographic plate. The separation of the centriods of the two images was a measure of the "blueness" of the stellar spectrum. In a paper he submitted in January 1915 entitled *Effective Wave-lengths of Absolutely Faint Stars*, he plotted his effective wavelengths against absolute magnitude. As Russell had found, the faint companion conspicuously stood out. Hertzsprung noted that "This exception is, in fact, very strange."

The next critical piece of evidence as to what might be going on was provided by Adams just a year later. He had been trying for two years to obtain a spectrum of Sirius B with the 60-inch Mt. Wilson reflector, then the world's largest telescope. He finally succeeded on October 18th, 1915, producing a spectrum of Sirius B that was reasonably free from contamination from Sirius A. He described the spectrum as being "identical to that of *Sirius* in all respects so far as can be judged from a close comparison of the spectra." As Adams pointed out, there was no question regarding the mass of Sirius B being nearly equal to that of the sun. In short, 40 Eridani B was not a fluke, since now there were two faint "A stars" occupying the same region of Russell's plot and one clearly possessed the mass of the sun.

The "A" star classification of the spectra of 40 Eridani B and Sirius B went unchallenged until 1922 when the Swedish astronomer Bertil Lindblad (1895–1965) obtained a detailed spectrum of 40 Eridani B and noted a number of peculiarities that did not fit with the spectra of any known A star. In particular he found the faint star showed the hydrogen lines (which is all the spectrum shows) to be abnormally broad and that the sequence of hydrogen lines seemed to weaken and fade out at its seventh member, the "η" line (the transition between energy levels 9 and 2 in the hydrogen atom), and further there seemed to be faint continuum at wavelengths shorter than the limit of the hydrogen series. In a normal A star the hydrogen lines are sharp, deep, and narrow and they march in a tightening sequence up to the limit of the hydrogen series at 3646 Å, at which point the intensity of the spectrum of a normal A star falls dramatically. None of this exactly fit with the "A" star classification of 40 Eridani B, but no explanation was forthcoming at the time.

It was Arthur Eddington who most clearly saw the astronomical and physical implications of these peculiar new stars. In his famous 1924 paper *On the Relation between Masses and Luminosities of the Stars*, Eddington having successfully dealt with giants and dwarfs took up the problem posed by Sirius B and the other strange faint "A" stars. In a brief section, somewhat separate from the rest of the paper, he comments on the extreme density of Sirius B. He used his theory of stellar interiors to point out that such small dense stars could form if the matter in the interior was fully ionized so that individual atoms completely dissolved into a sea of naked nuclei and free electrons.

The resulting separation of the electrons and nuclei would allow a very large compression of the matter, since atoms are mostly empty space. It was characteristic of Eddington that having satisfied himself that an explanation of Sirius' low luminosity was at hand, he proceeded further. In one of his more famous paragraphs he elaborates on the thermodynamical dilemma that such a fully ionized star would imply:

> "I do not see how a star which has once got into this compressed condition is ever going to get out of it. So far as we know the close packing of matter is only possible so long as the temperature is great enough to ionize the material. When the star cools down and regains the normal density ordinarily associated with solids, it must expand and do work against gravity. *The star will need energy in order to cool.* We can scarcely credit the star with sufficient foresight to retain more than 90 per cent in reserve for the difficulty awaiting it. It would seem that the star will be in an awkward predicament when its supply of sub-atomic energy ultimately fails. Imagine a body continually losing heat energy but with insufficient energy to grow cold!"

In a series of popular lectures given in 1926, and later developed into his book *Stars and Atoms*, Eddington again displayed his insight by marveling at the very different fates of Sirius A and B. The two stars must be of the same age and of basically the same initial composition. Therefore, why are they so remarkably different? "I feel convinced that there is something of fundamental importance that remains to be discovered," he wrote.

For sometime after the realization of the low luminosities of Sirius B and 40 Eridani B, no additional examples of these curious subluminous stars emerged. Then a young Dutch astronomer named Adriaan Van Maanen, working at Mt. Wilson during the First World War, began seeking companions of stars which showed very large proper motions. He was following a strong Dutch astronomical tradition of the statistical study of the positions, motions, and luminosities of the stars. Van Maanen compared photographic plates of the fields surrounding fast-moving stars taken in 1917 with previous plates of the same fields obtained in 1914. Quite by accident he noticed an unrelated rapidly moving star on one pair of plates. This 12.9 magnitude star was moving at a very high rate of 3 arc seconds per year, which implied that it must be quite close. He obtained a spectrum of the star with the 60-inch reflector and found its spectral type to be "F", which would have made it somewhat hotter than the sun. This indicated that the star was almost certainly subluminous, like Sirius B and 40 Eridani B. From its proper motion he estimated a distance of 1.4 pc (the modern value is 4.3 pc). For a time the star lacked a formal designation and was simply called "Van Maanen's F star"; today it is known as "Van Maanen 2". Now there were three of these strange stars. In 1922 Hertzsprung called attention to the fact that all three of these odd stars were quite nearby. The conclusion he drew from this observation was these stars were not rare subluminous anomalies but must in fact be the representatives of a larger population of such stars and common inhabitants of the solar neighborhood.

The discovery of Van Maanen 2 indicated a fruitful new way to search for additional stars of this type: look for faint, rapidly moving stars with spectral types A or F. It was another young Dutch astronomer, Willem Luyten, who took the next steps forward. Luyten had just finished his Ph.D. as Herzsprung's first graduate student at Leiden and had recently joined the staff of Lick Observatory at Mt. Hamilton, California. In 1922, at the age of 23, Luyten (Figure 7.4) published

Figure 7.4. Willem Luyten in 1925 (*My First 72 Years of Astronomical Research: Reminiscences of an Astronomical Curmudgeon*, W. Luyten, 1987).

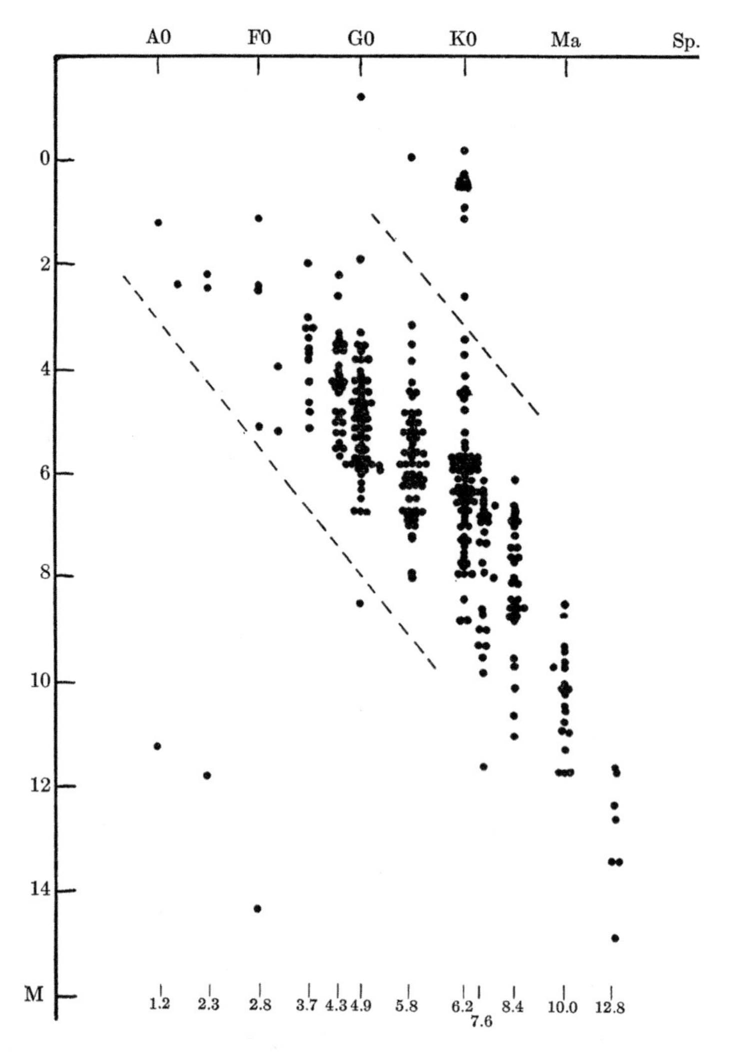

Figure 7.5. Luyten's 1922 plot of absolute magnitude vs. spectral type for stars near the sun. The three white dwarfs known at the time (Sirius B, 40 Eri B and Van Maanen 2) are located in the lower left-hand corner (*Lick Observatory Bulletin*, 1923, No. 344).

a series of three short notices in the *Publications of the Astronomical Society of the Pacific*, pointing out other possible instances of such stars. In the first note in January, he refers to these stars as "faint white stars". In the second note, in March, he describes them as "white stars with low apparent magnitude and large proper motion", and in the last note in November, he finally settles on the shorter more pithy term "white dwarfs". This is the name that finally stuck. He had actually first used the term "white dwarfs" to describe Sirius B, 40 Eridani B, and Van Maanen 2 in a catalog of the high proper motion stars that he completed in June of 1922, but which did not appear in the *Lick Observatory Bulletin* until 1923 (Figure 7.5). In retrospect, it is ironic that none of the stars Luyten called attention to in his 1922 *Publications the Astronomical Society of the Pacific* notes, is actually a white dwarf, rather they are for the most part, high proper motion A stars. Nevertheless, Luyten's work and his snappy descriptive label for these new stars attracted attention in important places. Eddington freely used the phrase *White Dwarf* in his 1924 mass–luminosity paper and it was Eddington's frequent use of the term shortly thereafter that popularized the term "white dwarf", forever fixing it in the astronomical lexicon. The expression Luyten coined in 1922 finally caught up with him in 1970 when he submitted a small grant to the National Science Foundation to help support a conference on "white dwarfs" to be held in St. Andrews, Scotland. After a few weeks, he received a letter from the United States Surgeon General demanding to know if human subjects were being used in experiments at the conference and reminding him that Federal funds cannot be used for segregated conferences. This bureaucratic obtuseness was a source of great amusement to Luyten.

Luyten had found his life's calling during his brief stay at Lick Observatory. He embarked on a monumental search of the heavens for additional high proper motion stars, in order to study the types of stars that are the sun's closest neighbors in space. His technique was that of Van Maanen, only on a vastly larger scale. He obtained thousands of plates of the sky at different epochs and systematically compared them for stars which moved with respect to the host of fixed stars. In this effort, he succeeded in discovering hundreds of genuine white dwarfs. Luyten used a blink comparator, a clever way of projecting two plates of an identical area of the sky taken at different epochs with the same telescope. In the blink comparator these two images were superimposed so that each star was precisely registered with itself on both plates, then the view was rapidly switched between images. Stars showing no relative motion appeared stationary, however, any star which showed the slightest shift between plates appeared to "dance" back and forth. Among a field of thousands of fixed stars, the human eye is particularly adept at noticing such minute motions.

Luyten spent the next six decades measuring plates in this manner. His output was phenomenal. Although blind in one eye, due to an early tennis accident, Luyten estimates he examined some 60 million stars, distilling them down to a sample of more than 100,000 which show varying degrees of proper motion. The sample of stars showing the largest proper motions, of one-half second of arc per annum and greater, are designated the "LHS" stars, for "Luyten Half arc-Second stars". Sirius itself, although obviously not discovered by Luyten, easily meets this criteria and is designated LHS 241 (Sirius B is LHS 241B). The LHS stars (numbering about 4000) contain the bulk of the population of the closest stars to the sun. Although most are not white dwarfs, many are. His largest sample, which exceeds ten thousand, is the "LTT" (for "Luyten Two Tenths") survey, which have proper motions which exceed 0.2 arc second per year. Luyten spent most of his career, until his death in 1994, at the University of Minnesota, measuring plates and publishing lengthy catalogs. He was an unusually blunt and outspoken man, but those who knew him and worked with him knew him as remarkably kind and generous. Astronomers who published follow-up observations on "Luyten stars" were well advised to use the proper Luyten catalog designation in referring to any star he had first discovered, or risk a wrathful letter from Willem Luyten pointing out the omission. As a testament to Luyten's success, approximately 25% of a recent catalog of 3300 known white dwarfs published in 1999 owe their initial discovery to his efforts.

For many years high proper motion coupled with a color bluer that the sun was one of best ways of finding white dwarfs. By the end of the Second World War there were approximately 80 confirmed white dwarfs. A few years later, a new technique emerged. Milton Humason and Fritz Zwicky, working at Mt. Palomar in California, began taking photographic plates in red and blue light of uncrowded regions of the sky where few luminous blue stars were to be expected. They found a number of faint blue stars, many of which turned out to be white dwarfs. This prompted a host of other similar surveys which have turned up the bulk of the known white dwarfs, along with a host of other peculiar blue stars, and distant quasars, located well beyond our galaxy.

Today most white dwarfs are being discovered as a byproduct of huge surveys of the sky with sensitive electronic cameras designed to hunt for faint galaxies and quasars. Many white dwarfs have colors that can mimic such faint extragalactic objects, therefore when these color-selected objects are subjected to later spectroscopic study, a small fraction are found to be new white dwarfs. Recently one such effort, the Sloan Digital Sky Survey, which has imaged nearly a quarter of the sky in the northern hemisphere, has already found nearly as many new white dwarfs as were discovered in the previous 90 years; and this is from only the first 20% of the full survey area. Spectroscopically verified white dwarfs and their properties are cataloged in the *Villanova Catalog of White Dwarfs* (*http://www.astronomy.villanova.edu/WDCatalog/index.html*) and at the *White Dwarf Database* (*http://procyon.lpl.arizona.edu/WD/*).

One type of white dwarf that is difficult to find with the above techniques consists of stars like Sirius B, where the existence of such a faint star is effectively hidden by the glare of a much brighter companion. The discovery during the 19th century of Sirius B, described in this book, is actually a remarkable accident. Had Sirius been ten times farther away it is doubtful the existence of the white dwarf would have ever come to light until the space age. First, if Sirius were ten times more distant, it would be a 3rd magnitude star: unremarkable but still easily seen by the naked eye. Its faint companion would only achieve a separation of 1″ every fifty years. It would be an almost impossible target in all but the most modern and specially adapted telescopes. Its existence might have been deduced from the proper motion variations (long after Bessel first did this for Sirius A in 1844), or perhaps from radial velocity variations in Sirius A. However, it is entirely possible that we would have been unaware of its existence until the early 1990s when the first satellite surveys of the sky were conducted at extreme ultraviolet wavelengths (see Chapter 11). Indeed only a handful of Sirius-like systems, close binaries, consisting of a bright main sequence star and a hot white dwarf were ever discovered by traditional ground-based means.

Two satellites, the *Extreme Ultraviolet Explorer* (*EUVE*) and *ROSAT* (*The Roentgen Satellite*), conducted surveys of the entire sky at very short wavelengths in the range between 100 Å and 700 Å—the extreme ultraviolet or EUV band. Such radiation simply cannot penetrate the earth's atmosphere and even has difficulty traveling very far in our own galaxy before being absorbed by interstellar hydrogen.

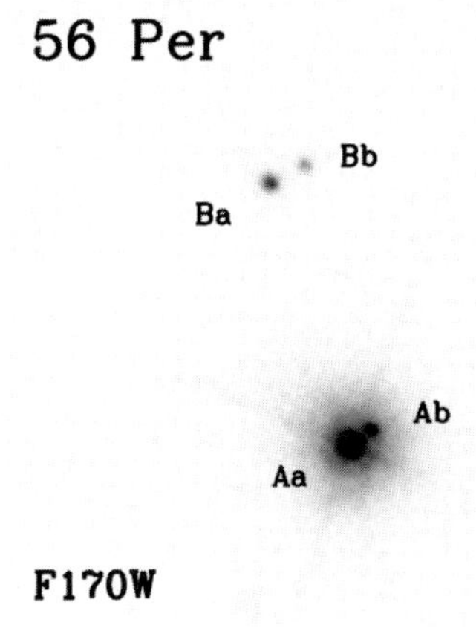

Figure 7.6. The Sirius-like star system 56 Perseus as seen by the *Hubble Space Telescope* Wide Field Camera. 56 Per consists of four stars. The primary is labeled "Aa" and the white dwarf is labeled "Ab". The white dwarf is estimated to have an orbital period of 47 years (from Barstow *et al.*, 2001).

Nevertheless, hot white dwarfs such as Sirius B radiate fiercely in the EUV and will even outshine huge luminous A stars like Sirius A at these wavelengths. One of the peculiarities soon noticed when these EUV satellites began observing was that not only were many known and unknown hot white dwarfs being detected in the EUV but that a number of seemingly common stars of spectral types A to K were abnormally bright at EUV wavelengths. The simple answer, in most cases, was that these stars harbored hidden hot white dwarf companions whose existence had been previously unexpected. About thirty such systems were found in this way.

Once discovered there were several ways to verify the existence of such Sirius-like systems. Observations can be conducted by other satellites in the ultraviolet and the telltale spectrum of the white dwarf might be seen. Indeed, the *EUVE* satellite itself had a spectroscopic mode that allowed spectra of the suspected stars to be obtained at EUV wavelengths, where the white dwarf spectrum could be easily spotted. Such observations not only confirmed the existence of the white dwarf but also permitted estimates of the temperature of these stars. In all, we now know of approximately 51 Sirius-like systems, most discovered in the EUV. It is hoped that surveys of the ultraviolet sky from space will reveal many more.

Careful images of 13 of these systems were made in 2001 with the *Hubble Space Telescope* cameras. These observations allowed a very close look at the suspect stars and actually obtained images of the white dwarf in nine of the cases (Figure 7.6). Once the white dwarf can be imaged it is possible to estimate the angular separation of the two stars and estimate the orbital periods. Periods of 50 years to as much as 1500 years were found in the instances where the white dwarf components were imaged. In those instances where no white dwarf was seen the separation between the companion and white dwarf must be much closer and, due to Kepler's Third Law, the corresponding orbital periods are shorter.

8

A Matter of Degeneracy

"The white dwarf appears to be the happy hunting ground for the most revolutionary developments of theoretical physics"
—Sir Arthur Eddington, 1926

Eddington summarized the dilemma concerning white dwarfs in one of a series of popular talks given during the summer of 1926 at King's College, London. These talks were later included as part of his popular, widely read book *Stars and Atoms*. One chapter, cast in the form of a detective story, was entitled *The Nonsensical Message*, by which Eddington meant that the messages (observations) astronomers had received from Sirius B seemed to make no sense. He recounted the history of Sirius from prehistoric Egypt through Bessel, to the discovery of the faint companion during the American Civil War. Eddington then described how the observations of Walter Adams, first of the spectral type and the high temperature of Sirius B and then of its gravitational redshift (see Chapter 9), inescapably led to the conclusion that the star must be incredibly compact and dense; as he put it, made of a "*material* 2,000 times denser than platinum." "It is scarcely possible to calculate what is the condition of the material in the Companion of Sirius, but I do not expect it to be a perfect gas," he wrote. He concluded, "I have told the detective story so far as it has yet unrolled itself. I do not know whether we reached the last chapter."

The resolution of Eddington's Sirius mystery, but not its last chapter, soon arrived in the form of a paper entitled simply *On Dense Matter*. Although the paper was read by Eddington at the December 1926 meeting of the Royal Astronomical Society, the author was an athletic 37-year-old Cambridge mathematical physicist, and wounded veteran of the First World War's Gallipoli campaign, Ralph Fowler. Fowler (1889–1944) had the insight to realize that the new theory of quantum statistics, independently developed just months earlier by Enrico Fermi in Italy and Paul Dirac, in Cambridge, had a direct celestial application to the interiors of dense stars. Fowler's interest in the topic was far from incidental. His mathematical career at Cambridge had taken him into the applied areas of thermodynamics and statistical mechanics. Indeed, Dirac had studied under Fowler at Cambridge, and it was Fowler who introduced Dirac to quantum mechanics. In his short 8-page paper, Fowler asked and answered a fundamental question about the nature of the interiors of compact stars. What are the thermodynamical properties of a dense collection of electrons and bare nuclei in a compact stellar core? Fowler found that the bare nuclei and free electrons played two opposing roles inside such a star. It was the nuclei that contributed essentially all the mass and were thus responsible for the crushing gravitational fields. On the other hand, it was the free electrons that were responsible for most of the internal energy and the pressure that balanced these gravitational forces. The electrons behaved very differently from the ordinary gaseous atomic matter in these stars. In particular, electrons did not obey the familiar gas laws with respect to temperature and pressure. Indeed, Fowler found that the very concept of temperature itself became almost irrelevant in such circumstances.

In 1926 the physics that governed the behavior of a collection of free electrons was virgin territory, ripe for exploration. In exploiting this new domain the recently discovered laws of the "new" quantum mechanics were paramount, in particular the Pauli Exclusion Principle. Named after the Austrian physicist Wolfgang Pauli, who proposed it in 1925, the principle states that no two electrons (or any particles with the same spin properties as electrons) can have precisely the same set of quantum numbers. The principle was conceived to explain why electrons fill the discrete energy levels in an atom in an orderly hierarchal manner and don't all collect together in the lowest energy level. It has come to be recognized as a fundamental property of electrons and other particles which have spin properties similar to an electron. Pauli's principle allows each discrete energy level in an atom to be filled by a single electron. Actually in an atom, pairs of electrons, one with clockwise spin and the other with counter-clockwise spin, occupy each of the atom's major energy levels, but on closer examination these major energy levels are found to be finely split so that the two paired electrons with opposite spins each occupy slightly different energy levels, since their spin quantum numbers differ.

The Pauli principle is not restricted to electrons in atoms but it also governs how free electrons occupy available energy levels in the dense core of a star. Fowler imagined a vast sea of free electrons flooding the deep well of potential energy in a dense stellar core. According to the Pauli Exclusion Principle the electrons would occupy and fill all available energy levels up to a certain point allowed by the set of possible electron quantum numbers. The statistics developed by Fermi and Dirac give a particularly simple picture of this electron sea. There is a maximum energy level, and a corresponding maximum level of momentum, for the electrons and all available levels of energy and momentum below these levels are fully filled and those above these levels are unoccupied. When Fowler worked out the consequences of these ideas, he found that it was the quantum mechanical pressure of the electrons and not the massive nuclei that checked the forces of gravity and prevented the collapse of the star. A stable star was therefore possible and the core of such a star was composed of a new type of material, degenerate matter. The term degenerate confers no moral judgment on white dwarf stars, rather it derives from a long-standing mathematical and physical term for situations in which distinct solutions to equations can merge into a single solution.

Fowler's paper marked the first direct application of quantum mechanics to stellar interiors. Although he discussed no particular stars, such as Sirius B or 40 Eridani B, beyond the need to consider the extremely high densities, he made one absolutely fundamental contribution to white dwarfs: he specified the equation of state for these stars. In nondegenerate stars, such as the sun, the equation of state, the proportional relation between the pressure (P), the density (ρ), and the temperature (T) of a gas is the familiar $P \propto \rho T$. As Eddington had shown, even in the core of the sun, where densities reach many times that of lead, matter obeys this relation and behaves much like a gas in the laboratory. For degenerate stars, temperature becomes largely irrelevant and the corresponding equation of state is $P \propto \rho^{5/3}$. Fowler did not explore the consequences of his ideas beyond the physical state of the matter in white dwarfs, and in particular he did not use his equation of state to build a mathematical model of an actual white dwarf, something that was relatively straightforward using standard polytrope models.

Fowler's paper on dense matter not only resolved Eddington's immediate problem with dense stars but it also set in motion a chain of events that would eventually embroil Eddington in the most famous of the many controversies of his professional life and also launch the career of another most unlikely astronomer.

In 1929 a promising young undergraduate named Subramanyan Chandrasekhar was finishing his studies at Presidency College in Madras, India. He had been born in 1910 into a well-to-do and remarkable Tamil family that would produce two Nobel Physics Prize winners in two generations. His father, C. S. Ayyar, was a high-ranking accountant in the Indian Railway System, who in his spare time played the violin, pursued the serious study of southern Indian music, and read English classics. When Chandrasekhar, or "Chandra", as he has come to be affectionately nicknamed, was seven the family moved to Madras where they occupied a large house filled with a library accumulated by Chandra's grandfather, who although largely self-educated, became a university official. It was in this house, surrounded by books in English and Tamil, that Chandra and his siblings were educated by both parents until old enough to enter high school.

As a teenager Chandra had decided on a career in pure research in mathematics or physics, rather than joining the Indian Civil Service as his father wished. On his own in 1927 Chandra had acquired and mastered an English-language copy of *Atomic Structure and Spectral Lines*, the German physicist Arnold Sommerfeld's authoritative exposition of what was known of the subject of atomic spectra in the early 1920s. This book was the high-water mark of the "old" quantum mechanics of the atom and was a difficult text for a second-year undergraduate. Physics was more than a remote academic subject for Chandra: his uncle was C. V. Raman, who had earlier made a fundamental discovery in molecular physics, which would lead to India's first Nobel Physics prize in 1930. In the summer of 1928 Chandra had traveled to Calcutta to work in his uncle's research laboratory. His main achievement that summer was managing to break an expensive piece of equipment. This incident convinced him that if he were ever to do research, it must be as a theorist. During the fall term back in Madras, in a remarkable coincidence, Arnold Sommerfeld visited Madras and spoke at Presidency College. After the talk, Chandra called on Sommerfeld at his hotel and expressed his interest in physics. Sommerfeld invited him back the next day and listened as Chandra proudly described his mastery of Sommerfeld's book. Sommerfeld frankly told the young man that all his studies in atomic physics were sadly out of date: soon after the book had appeared, an entirely new interpretation of the atom and quantum mechanics had been developed by individuals such as Werner Heisenberg and Paul Dirac. Sommerfeld, seeking to raise the boy's spirits, then asked what else Chandra had studied and Chandra replied that he knew something of statistical mechanics. Sommerfeld then handed Chandra the galley proofs for a soon-to-be published paper on the quantum behavior of the conduction (free) electrons in metals. The paper contained enough of a discussion of the new Fermi–Dirac statistics that within two months Chandra was able to produce an original paper on the application of the new statistics to the process which describes the interaction of X-rays and electrons, the Compton Effect.

Buoyed by his success, Chandra was searching for additional problems in which to apply the new statistical theory, when he came across Ralph Fowler's paper on dense stellar matter among some recently arrived journals. Chandra sent Fowler a copy of his Compton Effect paper, seeking his opinion on whether it was publishable. Fowler looked over the paper, and gave it to a colleague, Arthur Edward Milne (1896–1950), who provided a favorable judgment. Fowler wrote back saying he would gladly recommend publication, subject to a few minor changes. In the months it took to exchange sea mail between Madras and Cambridge, Chandra had made further progress on his paper as well as finishing a second paper on ionization in stellar interiors, with particular emphasis on Sirius B. On Fowler's recommendation both papers were to be published in prestigious British science journals. In the fall of 1929, during Chandra's final year at Presidency College, a second distinguished visitor arrived, Werner Heisenberg, the 27-year-old German developer of the "new" matrix mechanics approach to quantum theory. As the secretary of the science students, it was Chandra's honor and responsibility to accompany Heisenberg on a sightseeing tour of Madras. Chandra was delighted by the opportunity to meet with Heisenberg, who at the time was the leading physicist in Europe, next to Niels Bohr.

Chandra graduated with top honors from Presidency College in May 1930. In the normal course of events he would have taken an examination for the limited number of scholarship awards at universities in Britain. However, it was too late to sit for that year's exams and the Madras university officials, realizing that they had a very unusual student on their hands, created a new scholarship, particularly for Chandra. His examination was a mere formality. With the prospect of a scholarship in hand, Chandra again wrote to Ralph Fowler, this time requesting permission to study physics under him at Cambridge. Fowler agreed and arrangements were made. In July of 1930 Chandra traveled with his family from Madras to Bombay and arranged passage to Europe. During his week in Bombay, Chandra put the final touches on the third paper he had written while still an undergraduate. This paper took the obvious

Figure 8.1. Subramanyan Chandrasekhar in 1939 (Yerkes Observatory photograph).

next step following Fowler's work, the calculation of the internal structure of a white dwarf star.

On July 31, 1930, amid a throng of family well-wishers who had traveled to see him off, the 19-year-old departed Bombay on a steamer bound for Venice. The celebrations were muted, however, because his mother Sitalakshmi was too seriously ill to make the trip to Bombay. The Indian Ocean crossing was far from pleasant. Chandra found himself suffering seasickness while trying to adjust to European cuisine and still maintain his strict vegetarian diet. After the ship passed Aden and entered the Red Sea, the voyage settled into a more relaxed routine and Chandra recovered enough motivation to sit on deck and turn his mind back to physics. He had with him the paper he had written in India, which extended Fowler's work. In it Chandra had computed the structure of a white dwarf using Fowler's equation of state and standard stellar polytrope models. As he considered the conditions near the center of the star, he began to wonder about the effects of Special Relativity (Appendix A).

Chandra (Figure 8.1) noticed that the energies of the electrons in the core of a white dwarf could approach, and even exceed, the relativistic rest energy of an electron, given by Einstein's equation $E = mc^2$. This was a sure sign that relativistic effects needed to be included. Fowler had only considered nonrelativistic electrons. What would happen if special relativity were taken into account? Chandra had, in his baggage, the tools he needed to decide the matter, including Sommerfeld's book,

Compton's book on X-rays and electrons, and Eddington's book on stellar structure. To Chandra's surprise the answer he got was that in the extreme relativistic limit, degenerate matter took on a significantly different equation of state. The pressure varied as the 4/3 power of the density, $P \propto \rho^{4/3}$, not as the 5/3 power ($P \propto \rho^{5/3}$) as in the nonrelativistic case. It was well known that stars corresponding to polytrope models with this type of pressure dependence were not stable. The meaning of this instability was that there seemed to be a limit to the mass of a white dwarf and any star exceeding this mass would shrink without apparent limit. To Chandra's further surprise the limiting stellar mass turned out to be completely specified by three fundamental constants of nature, Newton's gravitational constant, Planck's constant, and the speed of light, along with the mean molecular weight of the matter in the star. Fortunately, when Chandra calculated this mass limit he found that it turned out to be somewhat larger than the measured mass of Sirius B. Chandra's calculation had passed its first test: the white dwarf Sirius B was within the range of masses for which stability was predicted.

Chandra was confident of his calculations but uncertain as to the meaning of this strange limit. By the time he arrived in Venice, he had written up this startling new result as a brief second paper and was quite anxious to show both his papers to Fowler in Cambridge. After landing in Venice and taking a rail journey across the unfamiliar continent of Europe, Chandra arrived in London, only to find matters in a fine mess. There had been a series of miscommunications between Madras, the Indian High Commission, and Cambridge officials. It seemed Chandra lacked the proper letters and was only slotted for a master's degree program not a Ph.D. program; there was, it seemed, no possibility of attending Cambridge. Why didn't he simply enroll in University College, London, officials wondered? After all, that sort of qualification would be sufficient to guarantee a young colonial student a good position when he returned to India. Frustrated, Chandra wrote to Fowler but Fowler was off vacationing in Ireland and would not return until the start of the new term. Getting nowhere with implacable bureaucrats in London, Chandra traveled on his own to Cambridge to await Fowler's return. There, in between trips to plead his case before officials, he continued to work on the strange result he had discovered during his voyage.

Through dogged persistence, Chandra was finally admitted to Trinity College, Cambridge and rented a spartan two-room flat. He and Fowler finally met for the first time met on October 2, 1930, just before the beginning of term. As Fowler looked over Chandra's first paper, he nodded commenting, "Splendid." The second paper on the limiting mass for white dwarfs caused Fowler to pause, he indicated he wasn't quite sure what to make of it and would forward it to E. A. Milne at Oxford for his opinion. Chandra then settled into an isolated university life at Trinity College, attending lectures by Fowler on statistical mechanics, Dirac on quantum mechanics, and Eddington on relativity. He found Dirac to be serious and philosophical, while Eddington was often amusing and humorous.

Chandra's first white dwarf paper was promptly published in the *Philosophical Magazine*. However, in the process he was surprised to learn that he was not the only one to have noticed the need to consider special relativity in white dwarfs. A year earlier both Edmund Stoner at Leeds University in England and Wilhelm Anderson at

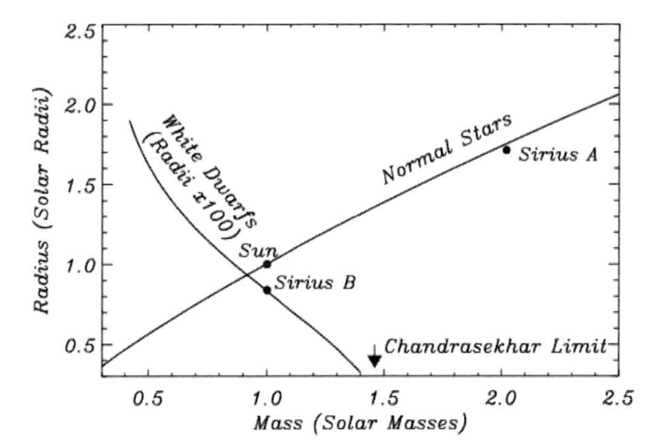

Figure 8.2. The different relationships between masses and radii for normal stars, like the sun, and white dwarfs. The white dwarf radii have been multiplied by 100 in order to place them on the same scale as the solar type stars. The locations of Sirius A, and B, and the sun are shown, as is the limiting Chandrasekhar Mass.

Dorpat in Germany had published results similar to those of Chandra, obtaining the same relativistic equation of state. Chandra dutifully acknowledged these results in his published paper. Stoner, a physicist, had only considered a hypothetical white dwarf of uniform density. Chandra, on the other hand, achieved somewhat more realistic results by considering how the density of matter changed with radius in polytrope models. Among his conclusions was that the radius of a white dwarf *decreased* as the cube root of the mass of the star. This seems to defy commonsense, the more massive the white dwarf, the smaller the star. On the familiar human scale, things like people and mountains generally *increase* in size as the cube root of the mass (Figure 8.2). For stars, like the sun, the radius increases at a rate closer to the square root of the mass, so that Sirius A, at slightly more than two times the mass of the sun, has a radius 1.7 times larger. White dwarfs were clearly objects with exceptional properties.

Chandra's second paper seemed to be languishing with Milne, who failed to grasp the significance of the limiting mass. Without the approval of Milne, or someone equivalent, it was unlikely the paper would be published in England. Out of frustration Chandra finally sent the paper to the *Astrophysical Journal* in the United States, where it was published in the spring of 1931, after some revisions. It is this short two-page paper that contains Chandra's famous result on the limiting mass. In the paper he calculated that no white dwarf could form with a mass greater than 0.91 solar masses. Sirius B in the 1930s was reckoned to have a mass of 0.85 solar masses, just under this limit, but not by much. The modern determinations of both numbers are 1.44 solar masses for what has now become known as the Chandrasekhar Mass or the Chandrasekhar Limit, and 1.0 solar masses for Sirius B. Chandra's lower estimate of the limiting mass is due to his choice of 2.5 atomic mass units for the mean molecular mass of the material in a white dwarf, such as Sirius B. A more realistic estimate for the mean molecular mass in the core of a white dwarf, such as Sirius B, is now known to be closer to 2.0 atomic mass units.

Figure 8.3. Sir Arthur Stanley Eddington in 1931 (permission Science & Society Picture Library).

At Trinity College, Chandra was both personally and intellectually very much on his own. He made only a few friends among the Cambridge students, devoting most of his time to his studies and his research. He saw Fowler infrequently and eventually Dirac became his official advisor, after Fowler left on sabbatical in the second term. Dirac, however, was not much interested in the astrophysics that Chandra was pursuing. Most of Chandra's intellectual contact was with Milne at Oxford, who suggested a number of collaborations in the somewhat unrelated field of stellar atmospheres. During this period Chandra occasionally attended the monthly meetings of the Royal Astronomical Society at Burlington House in London as a guest of Fowler and Milne. It was at these meetings that Chandra witnessed Eddington in an arena very different from the Cambridge lecture halls.

Arthur Stanley Eddington (Figure 8.3) dominated early 20th century astrophysics like no one else. He had solved many of the key problems of stellar physics, and his 1926 book *The Internal Constitution of the Stars*, stood for many years as the standard treatise on the subject. By the age of 48 he had gained a knighthood, becoming Sir

Arthur. Eddington was also charming, witty, erudite, and supremely self-confident. One of the highlights of the monthly Royal Astronomical Society meetings of that era was the sharp and seemingly good-natured verbal sparring that went on between Eddington, James Jeans, and E. A. Milne. Eddington maintained long-running scientific disputes and deep intellectual rivalries with both men but was seldom bested during the spirited give and take that followed their exchanges. Chandra had his first face-to-face meeting with Eddington in May of 1931, when he was invited by Eddington to a meeting to discuss his work and to receive Eddington's personal congratulations on his many accomplishments during his first year at Cambridge. Unfortunately the meeting occurred on the very day Chandra received a wire from Madras informing him of the death of his mother, Sita. This news must have cast a pall over his encounter with Eddington.

In spite of his success and a steady output of papers, Chandra felt his academic career was not advancing as he had hoped and that his work was not receiving sufficient recognition. For a time he considered switching to the exciting new field of quantum mechanics or even to mathematics. During the summer of 1931 Chandra solicited and received an invitation from the physicist Max Born to spend a few months at Göttingen. There, Chandra could improve his German and attend Born's famous lectures on quantum mechanics. He greatly enjoyed this time away from Cambridge and his immersion into the intellectually intoxicating atmosphere of European physics during the early 1930s.

Chandra's agonizing over the path he had selected at Cambridge only deepened after his return from Göttingen. At Dirac's urging, Chandra spent the following fall term of 1932 in Copenhagen, the epicenter of European quantum mechanics, where Niels Bohr held court and the knights of the new order such as Heisenberg, Dirac and Pauli came to pledge fealty. Chandra found the atmosphere of Copenhagen, just 900 km to the east, very different from that of Cambridge. There he was suddenly among a number of up-and-coming physicists of his own age and socializing with his peers in a fashion he had never managed at Cambridge. He immediately took up a problem in theoretical physics, originally suggested by Dirac, concerning the statistics of similar particles. Within just a few weeks he believed he had arrived at a solution and submitted a paper to the *Proceedings of the Royal Society*. Dirac, however, sent him a brief note pointing out a fatal error. Chandra went back over his work and, to his chagrin, realized that Dirac was correct and subsequently he withdrew his paper. By December Chandra was again back at Cambridge working on problems in astrophysics, disappointed with himself at having failed to scale the Olympian heights of quantum mechanics. That spring, however, his spirits improved with an invitation from the University of Liege in Belgium to give a series of lectures in astrophysics.

Chandra returned to Cambridge from Liege and the continent in May of 1933. His Cambridge scholarship was coming to a close in August, after which time he would face the prospect of returning to India, to seek some sort of university research position. First, however, he would have to sit for his Ph.D. exam. His thesis, scheduled for June, would consist of some of his recent work on distorted polytropes. The examination took place in Eddington's study at Cambridge Observatory. Fowler arrived late, hurriedly pulling on his academic gown, while Eddington appeared in casual dress and slippers. Together they questioned Chandra for 40 minutes, with Fowler and Eddington often disagreeing with each other about Chandra's answers. Chandra learned a few days later that he had passed. Although personally disappointed with his nearly three years at Cambridge, Chandra was not completely resigned to returning to India. There remained a faint hope that he might obtain a fellowship at Trinity that would allow him a further three or four years to pursue research. Chandra raised this possibility with Fowler, who frankly told him his chances were not good. The fellowships, which were open to candidates of all disciplines, were very highly prized and keenly competitive. He was required to submit a written thesis and sit for two written exams. With little to lose, he made an application and took the exams in late September of 1933. Anticipating the inevitable, Chandra had made plans to leave Cambridge and spend his final months in England working with Milne at Oxford. On the morning of his last day at Trinity, he hired a taxi for the railway station and decided to make one final check of the bulletin board where the fellowship results were posted. To his amazement he saw his name among the successful candidates. He canceled the taxi and immediately sent a wire to Madras informing his father of his great excitement and joy. For his father the message meant that Chandra would not be returning home, any time soon.

The fellowship provided Chandra a new lease on life. He moved from his cramped student digs into more spacious quarters. He now enjoyed dining privileges at the Trinity high table. In his own way Chandra became very much the Cambridge man, acquiring habits of dress and manner that he would retain throughout his life. For the first time, he felt he actually belonged at Cambridge. He established friendships between other scientists and mathematicians, and for recreation he took lengthy Sunday walks to nearby towns and villages. He regained his self-confidence and worked to refine his theories of white dwarf interiors, systematically replacing his initial approximations with exact solutions. He also continued to work with Milne on stellar atmospheres. His reputation as a productive theoretician continued to grow and resulted in invitations that allowed him to travel abroad. Eddington now took an active interest in Chandra's work, all the time quietly hoping that Chandra's refinements would ultimately eliminate the troubling limiting mass. Chandra finished his labors toward the end of 1934, with the result that the limiting white dwarf mass had become more entrenched than ever.

Milne accepted Chandra's relativistic equation of state, but he disagreed with the conclusion that a white dwarf, which exceeded the limiting mass, might shrink without limit. Instinctively, he felt such a star would somehow achieve equilibrium at a finite radius. Chandra, Milne concluded, simply hadn't yet found the right answers. Eddington was equally opposed for many of the same reasons, but in addition he fundamentally rejected Chandra's relativistic equation of state. To resolve these issues, Chandra had earlier written a paper in which he carefully worked out a demonstration that the problem of the mass limit could not be avoided by a rearrangement of the mass within a star, as Milne had hoped. In the end, if a star were too massive, Fermi–Dirac statistics and electron degeneracy could not support the star. Chandra ended his paper with a fundamental question: "What happens," he asked, "if we go on compressing the material indefinitely?" The paper was published in the German journal *Zeitschrift für Astrophysik* in 1932 only after Milne reluctantly agreed to its acceptance. The sophisticated and detailed model that Chandra finally developed consisted of a core of fully degenerate matter containing most of the stellar mass. Surrounding this core was a zone of nondegenerate matter and finally a thin veneer, or atmosphere, of dense gas which emitted the observed light of the star.

Eddington stated his opposition to Chandrasekhar's limiting mass in basically philosophical terms. He had been relieved when Fowler rescued astronomy from the dilemma of a star which might contract without limit as it cooled. When Chandrasekhar introduced his relativistic equation of state, the unwelcome dilemma suddenly reappeared, this time for stars above a certain mass. Eddington instinctively felt that such a situation was absurd, stars would not behave in this manner, the problem must lie with the equation of state, but he couldn't identify exactly where the difficulties were. At the time Eddington was busy with many other matters and failed to object publicly to Chandra's results.

The inevitable collision between Chandra and Eddington finally occurred on January 11, 1935 at the monthly meeting of the Royal Astronomical Society. As was the practice, papers scheduled to be read were listed several days before the Friday of each monthly meeting. Chandra had submitted a paper refining his calculations on the cores of white dwarfs and was surprised to notice that he was to be followed by Eddington, speaking on the topic of "Relativistic Degeneracy". A few days before the meeting Chandra encountered Eddington at the Trinity College dining room and Eddington asked if his talk would be long. Chandra responded that it would and Eddington replied that he had asked the Society secretary to give Chandra 30 minutes for his talk, rather than the normal 15 minutes. Eddington revealed nothing of what he planned to say.

At the Society meeting at Burlington House in London, Chandra confidently gave his talk, in which he elaborated on his latest work on degenerate stellar cores and the limiting mass for white dwarfs. Afterwards, Milne stood up and made a few supporting remarks. Then Eddington took the podium and began by reviewing the recent history of white dwarf interiors, starting with his own realization that highly dense collapsed stars were possible and of the dilemma that these extreme stars had posed for astrophysics. He mentioned how the work of Fowler had resolved the dilemma of what prevents the collapse of such stars. This all led up to Eddington's appraisal of Chandra's relativistic equation of state and how for stars above a limiting mass, no stable configuration was possible. He remarked, "I do not know whether I shall escape from this meeting alive, but the point of my paper is that there is no such thing as relativistic degeneracy!" Returning to the paper that Chandra had just delivered and the limiting mass, Eddington described how "I felt driven to the conclusion that this was almost a *reductio ad absurdum* of the relativistic degeneracy formula. Various accidents may intervene to save a star but I want more protection than that. I think there should be a law of Nature to prevent a star from behaving in this absurd way!" Eddington acknowledged that the mathematics of Chandra's derivation of the relativistic equation of state was above suspicion, but went on to make the point that the quantum mechanics used was nonrelativistic, since Chandra had only introduced relativity through the physical relation between energy and momentum. Combining these two theories was suspect and inconsistent, Eddington argued and with characteristically colorful language he remarked "I do not regard the offspring of such a union as born in lawful wedlock." He ventured that when a fully relativistic theory was finally developed, Chandra's limiting mass would evaporate. Eddington followed this with an obscure distinction between the "standing" and "progressive" wave properties of electrons, using an amusing analogy involving the short portly President of Royal Society. After Eddington concluded his brief talk, the next speaker was promptly called, leaving no opportunity for discussion from the audience.

Chandra had been completely blindsided. Sir Arthur had just spoken from Mt. Olympus, and derisively brushed aside much of Chandra's last four years of work. Although the rhetorical style was vintage Eddington, the content of the talk was curiously devoid of much real substance. There was no explicit demonstration of why Chandra's relativistic equation of state was not appropriate and certainly no useful alternative was offered. Eddington's talk amounted to what physicists and mathematicians derisively refer to as "hand waving". Eddington's reputation commanded great respect and his opinions and views were regarded by most as the final word on a subject. Chandra, on the other hand, was a mere 24-year-old researcher, at the beginning of his career. After the meeting, friends approached Chandra and expressed their sympathy, saying "It was too bad, too bad." Chandra was stunned and hurt. Why had Eddington not revealed his views earlier, after all the two saw one another regularly? No one came to Chandra's defense and Chandra again began to doubt himself and his work. Out of frustration Chandra turned to the founders of quantum mechanics such as Niels Bohr, Wolfgang Pauli, and Paul Dirac, to seek their judgments. The physicists concurred with Chandra, and were unable to make any sense out of Eddington's arguments. Nevertheless, they were physicists and were consumed with discovering and arranging the fundamental building blocks of matter. They had little interest in the field of astrophysics, which at the time was unable to even answer the basic question of how the sun produced its energy. Ironically there already existed validation of Chandra from within the physics community. In 1932 the famous Russian physicist Lev Landau had derived Chandra's mass limit for a degenerate star using a particularly simple set of arguments, based on Pauli's Exclusion Principle and Heisenberg's 1927 Uncertainty Principle (Appendix A), which expresses the fact that there is a fundamental limit to our joint knowledge of the position and momentum of a particle. Landau's derivation of the mass limit neatly sidestepped the direct use of Fermi–Dirac statistics and relativistic equations of state and therefore would have been more difficult for Eddington to dismiss out of hand. Unfortunately this result never received as much attention as it should have at the time.

The dispute between Chandra and Eddington persisted into the summer of 1935, when Eddington made it the center point of a talk he gave at the International Astronomical Union. Chandra was in a difficult position since Eddington's judgment loomed like a dark cloud over his future. Before the open dispute with Eddington, Chandra had mapped out research plans for the next several years which involved extending his exact theory of white dwarfs to cases including rotation and stellar pulsations. It would be difficult to persist with these plans if he remained at

Cambridge. Given Eddington's reputation, it was unlikely Chandra's work would ever receive much serious attention, since most people would simply assume it was wrong.

A way out of this dilemma appeared in the fall of 1935, when Chandra received an invitation from Harlow Shapely, the director of the Harvard University Observatory in Boston, Massachusetts to give a series of lectures. Eddington might cast a long shadow in England and Europe, but in America, with its giant telescopes and mountain top observatories, it was observers and new ideas, not eminent Cambridge theoreticians, who commanded the most attention. Attracting a young and productive astrophysicist like Chandra had many advantages for American astronomy. To Chandra this invitation offered a temporary respite from Cambridge and an opportunity to consider his future. In addition to the difficulties with Eddington, there were also long-standing issues of family and marriage that beckoned to him from India. His father's letters continually inquired when he planned to return, take up a suitable position, and when he planned to marry. Chandra had considered several positions in India but regarded them either not suited to his talents or too closely involved with his uncle Raman's research institute in Bangalore. He secretly dreaded the prospect of finding himself in an Indian academic position where he would be saddled with administrative duties and be drawn into the political infighting that characterized much of Indian science during the period. Throughout his Cambridge years Chandra had also been corresponding with Lalitha, a young physics student whom he had met before leaving Madras. The two had established an "understanding" that they would marry, once Chandra returned to India. Chandra's family was relatively unique in that he had been allowed the latitude to choose his own bride, as had his uncle Raman and his grandfather, rather than the traditional Indian arranged marriage. Among his other doubts, Chandra was now having second thoughts about this long-distance relationship with Lalitha. It was in this atmosphere that he left for Harvard in early 1936.

His four-month Harvard stay proved very successful and resulted in the offer of a permanent position at Harvard as well as a counter-offer from the Yerkes Observatory, which was part of the University of Chicago. The Yerkes offer came from Otto Struve (1897–1963), the director of Yerkes and a member of the fourth generation of astronomers produced by the remarkable Struve family. On the return voyage to England, Chandra considered both offers, his future at Cambridge, and his plans to return to India and possible marriage. He decided that he would return to England, and then make a brief trip to India in the summer of 1936 to visit his family. He would call off his marriage plans with Lalitha and then take up the offer at Yerkes in January of 1937.

Chandra arrived in India in mid-August of 1936, for a short two-month visit. He had not seen his family in six years and during the brief stay there never seemed to be enough time to get reacquainted. He found himself a minor celebrity when he visited former friends and teachers at Presidency College and was pressed to give all manner of talks and lectures. His main purpose in returning was to finally resolve the issue of his marriage. Although he had earlier decided against marriage, when he met with Lalitha he changed his mind. They were married in an abbreviated Hindu ceremony one month after his arrival in India. A month after that, both Chandra and Lalitha were on a steamer headed back to England and then to America.

When Chandra arrived at Yerkes, he was given the task of building up a strong graduate program in astrophysics. During his first two years at Yerkes he devoted much time to writing his classic study of stellar structure, *An Introduction to the Study of Stellar Structure*. Published as a monograph in 1939, by the University of Chicago Press, the book was far from a mere introduction to the subject of stellar structure, it was a fully developed exposition of the state of knowledge in the field, just prior to the discovery of the nuclear reactions that power the stars. Much of the work in the book was drawn from Chandra's own contributions to the field and the rest bore the stamp of his thorough and complete treatment of topics. Chandra's book soon replaced Eddington's 1926 book *The Internal Constitution of the Stars*, as the standard reference in the field. Its publication also marked Chandra's withdrawal from the field of stellar structure. If he could not overcome Eddington, then Chandra thought it best to switch fields. He took up the theoretical study of the motions of stars.

As odd as it may seem, the dispute between Chandra and Eddington was never overtly bitter or personal. A good description is perhaps that of a superficially polite, but strictly enforced, Oxbridge intellectual "pecking order". Chandra, for his part, was hurt, frustrated, and reduced to venting his frustration in letters to his father. "Eddington is behaving in a most obscurantist fashion," he fumed in one letter and, "Eddington is simply stuck up!", he wrote in another. To Eddington it was a simple matter of just being able to seeing things his way. How could stars behave in such an absurd manner, he argued? It was as if Chandra had somehow managed to "prove" mathematically that the earth was flat; never mind the equations, essentially it must be nonsense.

Four years after the fateful January 1935 encounter, Chandra and Eddington met for the last time at a conference in Paris devoted in part to white dwarfs. At this conference the Dutch–American astronomer Gerard Kuiper presented some of the first observational evidence that seemed to favor the model of white dwarfs devised by Chandra. Chandra cited this evidence in his talk on development and consequences of relativistic degeneracy in stars. After Eddington's talk, Kuiper rose to ask if there were any observational tests which would permit a choice between the two rival theories (relativistic and non-relativistic degeneracy). Eddington replied that relativistic degeneracy did not exist and "observation can decide between rival hypotheses but not between rival conclusions which profess to represent the same hypothesis." Chandra then defended his relativistic formula and suggested that for the observer the two theories must be considered rivals. At the formal luncheon after the conference, Eddington approached Chandra and said, "I am sorry if I hurt you this morning. I hope you are not angry with what I said." Chandra curtly inquired, "you haven't changed your mind have you?" "No," Eddington replied. "What are you sorry about then?", Chandra responded and walked away. On the ocean voyage back to his exile in America, Chandra wrote to his father that he very much regretted his rudeness to Eddington.

Eddington was indeed stubborn, especially when it came to Chandra's discovery of the limiting mass. However, his objections were all quite reasonable from his point of view. In the early 1920s Eddington had been the chief architect who worked out the details of the internal structures of the stars. He had shown that it was mass that basically determined the luminosity of a star and he explained the range of masses in which stable stars could be found. He also was the person who realized that matter in most stars behaved largely as a gas and that this allowed the internal properties to be specified. When he encountered the peculiar internal properties of white dwarfs in 1924, he was at first puzzled, but when Fowler came up with the idea that the degenerate pressure of electrons could support these stars, he felt vindicated once again. In Eddington's mind, nature constructed stars that were stable and well-behaved; he had a comfortable familiarity with such stars and felt he instinctively understood how they were constituted. Unstable massive white dwarfs of the type Chandra was describing had no place in this scheme, they were simply impossible, and that was all there was to it.

Eddington died in Cambridge in 1944 at the age of 62, adhering to his views on white dwarfs to the very end. During the Second World War Eddington had maintained a lively correspondence with Chandra in America, interspersing scientific discussions with charming stories of Eddington's marathon bicycle rides through the English countryside and news of the wartime deprivations at Cambridge. It fell to Chandra to deliver the University of Chicago's obituary for Eddington. The last chapter in the dispute between Eddington and Chandra was finally written five decades later. In 1983 Chandra produced a short commemorative volume on the occasion of Cambridge's Centenary Lecture in honor of Eddington. The book, entitled *Eddington: The Most Distinguished Astrophysicist of This Time*, contained Chandra's assessment of Eddington's many contributions to astrophysics as well as Chandra's personal recollections of the man.

Ironically, in 1983 Chandrasekhar also shared the Nobel Prize in Physics with William Fowler, a nuclear physicist who helped work out how the chemical elements are formed in the stars. It is often the case that Nobel Prizes are given in recognition of work that was done decades earlier. In Chandra's case, however, the work cited by the Nobel Committee went back over five decades, to his discovery of the limiting mass of a white dwarf that he had made on his voyage to Cambridge in 1930. The official Nobel citation is rather general and mentions Chandra's work on "many aspects of the evolution of stars" and some of his "contributions to the study of the structure of white dwarfs". There was no explicit mention of his discovery of the limiting mass of a white dwarf or his masterful development of the theory of these stars. Noting the long time lag between his work and the award of the prize, the committee stated, "Even if some of these studies are from his earlier years, they have become topical again through advances in the fields of astronomy and space research." Many, however, felt that had it not been for his dispute with Eddington, the recognition would certainly have come much sooner. Chandra for his part was outwardly philosophical, noting ironically that he probably never would have received the Nobel Prize at all, were it not for the skills of the doctors who performed his heart surgery in 1975: a circumstance that allowed the Nobel Committee the extra eight years it required to adequately consider the issue.

Chandra never did return to India, except for occasional visits; he spent the rest of his life at the University of Chicago, first at Yerkes, and later moving to the University campus in Chicago. At Chicago he became known to all as the longtime editor of the *Astrophysical Journal*, the publication begun by Hale in 1895 to publish the results of the new field of astrophysics. In addition to his teaching duties and his editorship he managed to find time to continue his research. He was methodical and thorough in everything that he did and his research reflected that. After writing *An Introduction to the Study of Stellar Structure* in 1939, Chandra abandoned that field entirely and plunged into the study of stellar motions. Ten years later he summarized that field with another book, and then turned to his former field of stellar atmospheres. After another decade he summarized that field with a book, *Stellar Atmospheres*. All of these books are regarded by professional astronomers as classics in their respective areas. It was a pattern that he continued: he would identify a topic or field where he felt he could make a contribution and then he would plunge in and work intensively until he had mastered it and moved the existing body of research to a new level. He then would summarize everything in a comprehensive book and move on to a new area. Most of Chandra's work was highly mathematical but always presented with great clarity and insight. His last project, finished just before his death in 1993, was somewhat different, it involved a modern explanation of Isaac Newton's *Principia* in which he used modern notation and derivations to illustrate what Newton had achieved and to provide modern readers with insight into Newton's methods.

The question that Chandra had first posed in 1933, of what happens to degenerate matter when you go on compressing it, and the dilemma that bedeviled Eddington, of stars that became infinitesimally small as they cooled, had no obvious answer in 1935. The first theoretical inklings of an answer to these questions appeared later in the 1930s. The discovery of the neutron, a nuclear particle with the approximate mass of the proton, but neutral in charge, was made by James Chadwick at Cambridge in 1932. This led several people to investigate whether it might be possible to form stable stars from neutrons. Neutron stars, as they became known, were a theoretical possibility but they would be incredibly small, perhaps only about 10 km in radius and unbelievably dense. Another possibility that was investigated at the time was black holes. In 1939 it was shown that there was no physical reason why a very massive star could not shrink to such a small size that its gravitational field would become so large that no light could escape its surface. At the time, however, neutron stars and black holes were simply too strange and bizarre to be considered seriously. Moreover, no examples of such objects were known and, besides, it seemed they would be almost impossible to detect, even if they did exist.

It was only in the 1960s and later that the reality of such objects became unavoidable. The first neutron star was discovered by radio astronomers in 1967 as a rapidly spinning pulsar. As predicted, neutron stars turned out to be objects only about 10 km in diameter but having a mass greater than the sun. Over a thousand of these tiny rapidly spinning stars are now known. Neutron stars are the compact stellar analogs of white dwarfs and consist of a stellar core of degenerate neutrons rather than electrons. Neutron stars therefore also possess their own maximum mass, believed to be about 1.4 solar masses, which is analogous to the Chandrasekhar

Mass for white dwarfs. The bottom rung on the descending ladder of compact stellar objects is a black hole, which represents the fate of stars with cores more massive than the neutron star limit. Although no stellar black holes have yet been "seen" there is now plenty of evidence that they do exist. Observations at X-ray wavelengths reveal the final moments of matter being tortuously dragged down the throat of a black hole. Additionally, black holes which are members of binary systems have now had their masses measured. Calculations of the properties of black holes require Einstein's theory of General Relativity (see Chapter 9), which predicts that for massive bodies there is a threshold in space called the event horizon beyond which, not even light, can escape. The event horizon for a black hole having the mass of the sun is only about 6 km in diameter. Unlike white dwarfs and neutron stars, black holes have no theoretical mass limits. The giant black holes that inhabit the centers of some galaxies have masses that exceed a billion solar masses. When these huge black holes consume stars and gas from the surrounding galaxy they emit enormous quantities of energy and appear as quasars, the most luminous objects in the universe. Had the possibilities of neutron stars and black holes been more real at the time, Chandra's strange limiting mass for white dwarfs would not have been seen as the conundrum it was in the early 1930s.

In the 1930s only a handful of white dwarfs were known. Although the theory was in place that described the extreme conditions inside a white dwarf, no one at the time had any idea how nature produced these strange objects. The best idea at the time was that they were the natural product of stars which shrank as they cooled. The problem was that there was no good theory to explain how the star evolved and very little idea of how they produced their energy. The old idea that gravitational contraction powered stars had serious problems and had been largely abandoned. Eddington was one of the earliest and chief advocates of the idea that the new phenomena of radioactivity had to be involved, having first offered these ideas as early as 1920. A decade later the suspicion that stars obtained their energy by burning hydrogen into helium in their cores was widespread. It was, for example, possible to calculate the amount of energy released by the process of two hydrogen atoms fusing into helium. What was not known at the time was under what conditions these reactions occurred and at what rate they operated. The basic answer to these questions only became known in 1939, when Hans Bethe at Cornell University worked out the series of nuclear reactions that resulted in the net conversion of hydrogen into helium in the sun and the stars, and for which Bethe was awarded the 1967 Nobel Physics Prize.

The process by which white dwarfs are formed is now well established. Stars consume the store of hydrogen and helium in their cores, burning these elements to carbon and oxygen, and sometimes heavier elements. In the process they swell and shed their outer layers, leaving only a core of degenerate matter. The primary determining factor on the final outcome is the original mass of the star. Stars much less massive than the sun burn at such a low rate that they will only form white dwarfs after lifetimes several times longer than the age of the universe. The sun is half-way through its life as a hydrogen-burning star and will ultimately become a white dwarf composed of carbon and oxygen with a mass about 53% of its present value (see Chapter 13). Stars much more massive than the sun have significantly shorter lives.

The precursor of Sirius B originally had about 5 solar masses and the present remnant is only 1 solar mass. It is believed that stars as massive as 8 to perhaps as high as 12 times the mass of the sun will form massive white dwarfs very near the Chandrasekhar Limit of 1.4 solar masses. More massive stars cannot form stable degenerate cores and will explode in a type II supernova, when their supply of nuclear fuel is exhausted. In some cases a type II supernova leaves a remnant neutron star. Stars too massive to form neutron stars may well collapse into black holes.

One of the most important consequences of the white dwarf mass limit occurs when white dwarfs, which formed below the limiting mass, acquire sufficient mass to exceed the limit. This can occur in binary systems when a less massive companion gravitationally transfers mass onto the white dwarf, or perhaps when two white dwarfs merge and exceed the limit. In either case the result is a massive explosion, a type Ia supernova that completely disrupts the white dwarf. The suggestion that type Ia supernovas involve white dwarfs was first made by the Swiss astronomer Fritz Zwicky in 1935, shortly after Chandra's work on the limiting mass. Such events are inherently rare—we observe one approximately every three hundred years in our galaxy. The last two were the supernovas observed by Tycho Brahe in 1573 and Johannes Kepler in 1604. They are seen more frequently, however, if one patiently observes thousands of external galaxies. The extreme brightness of these distant events means that they can be observed all the way across the visible universe. The critical aspect of type Ia supernovas is that they all seem to explode with nearly identical energy so that they all appear to have the same intrinsic brightness. The reason for this similarity is that each explosion, no matter how it occurs, always involves a white dwarf with a core composed of carbon and oxygen nuclei that has exceeded the Chandrasekhar limit of 1.4 solar masses.

9

Einstein's Well

"I little imagined when this survey of Stars and Atoms was begun that it would end with a glimpse of a Star-Atom"

—Sir Arthur Eddington, *Stars and Atoms*, 1927

By the early 1920s Sirius B was poised to play another major role in the revolutions that transformed physics and astrophysics during the 20th century. The story begins in 1907 when Albert Einstein was still working as a clerk at the Patent Office in Bern, Switzerland. Just two years earlier he had astonished the world with five earth-shaking papers, including his Theory of Special Relativity that treated uniform relative motions. In 1907 he was taking his first steps towards developing a General Theory of Relativity, which involved gravitation and accelerated motions, when he published a paper that introduced his equivalence principle and first predicted that light emitted from a massive object would appear to be redshifted when viewed by a distant external observer. Four years later in 1911, Einstein—who by then had become a professor at the Karl-Ferdinand University in Prague—published a second paper in *Annalen der Physik* whose title in English is *On the Influence of Gravitation on the Propagation of Light*. Not satisfied with his earlier treatment of these subjects, Einstein reconsidered the consequences of his 1907 discussion of the equivalence principle. The equivalence principle, which was the genesis of his General Theory of Relativity, states that the influence of accelerated motion on a body, in the absence of a gravitational field, is effectively no different than the influence of a gravitational field on an unaccelerated body. This is why a freely falling body in a gravitational field experiences weightlessness. A consequence of the equivalence between gravitational acceleration and accelerated motion is that the mass used in Newton's law of universal gravitation, and the mass employed in Newton's second law of motion (Chapter 3) must refer to the same quantity. These two usages of the term "mass" had always been treated as if they were the same but few physicists had ever inquired deeply into the consequences of this equivalence. The question Einstein asked in 1911 was what happens if this principle of equivalence is extended to all physical laws in such a way that no conceivable experiment could be conducted which would distinguish between the effect of acceleration and the effect of a gravitational field? The way this fundamental principle is often framed is to imagine that you are in an enclosed elevator, then you would have no way of determining if you are sensing the earth's gravitational field, or if the earth's gravitational field had been magically turned off and the elevator was accelerating upward at 1 g. This is not simply a matter of deceiving the physiological senses but a result of the fact that all the laws of physics remain the same in either circumstance.

In 1911 the particular question Einstein considered was: What happens to light in a uniformly accelerating system? According to his 1905 Special Theory of Relativity, there is a direct relation between a body's mass and its energy described in the famous equation $E = mc^2$, where E is the energy, m is the mass, and c the speed of light. When a body's energy is increased, its mass increases proportionately. Therefore, Einstein asked: Would its gravitational mass also increase? He answered this question, as he often did, by using a simple but profound thought experiment. Einstein imagined two bodies separated by some fixed vertical distance in a uniform gravitational field. The bodies exchange energy by emitting and absorbing radiation (light), effectively decreasing and increasing their mass by an amount $\Delta m = \Delta E/c^2$, where Δm and ΔE are the corresponding changes in mass and energy. He then considered the same two bodies, but this time exchanging energy while uniformly accelerating. Using simple arguments that can be followed by most undergraduate physics students, such as the Doppler shift and the conservation of energy, he showed that the radiation itself had to gain or lose energy as photons were exchanged between the two bodies. He was further able to calculate the amount of this change as a function of the acceleration and the distance between the bodies. Merely replacing the acceleration with a gravitational field gave Einstein a simple expression for how light loses energy as it escapes from a massive body. One of the five epochal papers Einstein wrote in the year 1905 had already introduced the concept that quanta of light possess an energy that is proportional to its frequency $E = h\nu$, where ν is the photon's frequency and h is Planck's constant. From this relation the loss of energy of light escaping from a massive body is reflected in a proportional decrease in frequency, which in turn corresponds to an increase in its wavelength λ ($\lambda = c/\nu$). This is the gravitational redshift. When light of a given wavelength is emitted at the surface of a massive body its wavelength is stretched so that the light appears redder when observed far from the body.

An equivalent way of thinking about the gravitational redshift is that a clock will run slower deep within a gravitational field, with respect to a clock far outside of the gravitational field. The "depth" of the gravitational field is measured in terms of its gravitational potential, or "gravitational potential well", which is defined by the ratio of the mass of a body to the distance from its center. If the clock is an oscillating atom emitting light at a characteristic frequency, then the frequency of the light will be lower with respect to light emitted by a similar atom far from the source of the gravitational field. Effectively photons of light lose energy climbing out of the gravitational potential well of a massive body. The corresponding lengthening of the wavelength behaves precisely like a Doppler shift and the wavelength change is often expressed as an equivalent Doppler shift velocity, even though no relative motion is involved.

In 1911, the massive body that Einstein originally had in mind to demonstrate this effect was the sun. It is simple to calculate what the gravitational redshift velocity ought to be for photons emitted from the surface of the sun and observed at the earth: 6/10 of a kilometer per second. Although such a velocity is equivalent to twice the speed of a commercial jet airliner, it was a difficult matter to unambiguously measure the Doppler shifts of this magnitude using the instruments of Einstein's day. Einstein cited some observations that appeared to show such a Doppler shift for Fraunhofer absorption lines in the sun's spectrum; however, the truth of the matter was that the observed effect was marginal at best. Even after several observers made repeated attempts to measure the sun's gravitational redshift, the issue remained equivocal.

It was ultimately a second result derived by Einstein in his 1911 paper that provided a more convincing verification of the predictions of General Relativity: this was the bending of starlight in the sun's gravitational field. In 1919 two British expeditions were sent to observe a total solar eclipse, one to the west coast of Africa and the other to Brazil. Both returned with photographic evidence that the apparent positions of the stars observed near the eclipse-darkened solar disk were shifted in the direction of the sun by the amount predicted by Einstein. Initially it was this bending of starlight, together with the precession of Mercury's orbit (see Chapter 4) that was cited as the primary evidence in support of the General Theory of Relativity. The 1911 paper represented one of a series of steps that eventually led Einstein to his complete theory of General Relativity which he published in 1915. It is this final comprehensive theory that introduces the ideas of curved space–time and contains the field equations that give rise to such phenomena as gravitational waves, black holes, and the expanding universe. What Einstein published in 1911 was much more elementary and did not rely on the sophisticated mathematical and theoretical framework that yielded these later concepts. It is also noteworthy that the original expression used by Einstein to calculate the bending of light was incorrect and underestimated the bending angle by a factor of 2, though Einstein had determined the correct expression long before the eclipse results were announced at a special joint meeting of the Royal Astronomical Society and the Royal Philosophical Society in November 1919. It was this confirmation of the Theory of General Relativity that earned Einstein his public status as the twentieth century's most recognized scientific celebrity.

In the 1920s it appeared as if a convincing observational demonstration of the gravitational redshift from a star would long remain a difficult feat. The magnitude of the gravitational redshift depends on the simple ratio of the mass to the radius of a star. Expressed in units of a solar mass (M) and a solar radius (R), the redshift velocity (V) in km/s is $V = 0.6M/R$. Even for stars more massive than the sun, the result never exceeds a few km/s, since the radius of such stars increases with mass in such a way that the ratio M/R changes relatively little. What was needed to produce a truly large effect was a very small, but massive star.

When Arthur Eddington had worked out his general relationship between the mass of a star and its luminosity, he was also able to estimate the star's temperature and its radius (see Chapter 7). He presented his ground-breaking results in a paper delivered to the Royal Astronomical Society in March of 1924 and demonstrated its effectiveness by showing that it produced results consistent with purely empirical measurements of bright stars. Then in a dramatic departure from the main theme of this paper, Eddington considered Sirius B and 40 Eridani B, noting that although they could not possibly fit his new relationship between mass and luminosity, these curious tiny stars could well be examples of the complete ionization of matter within a star. From its spectral type Eddington estimated that Sirius B must possess a temperature of 8000 K. From this and its absolute magnitude, he estimated its radius to be 19,600 km, "much less than Uranus". Although such a small radius implies a ridiculously large density, Eddington assured his audience, "According to the views here reached, such a density is not absurd, and we should accept it without demur if the evidence were sufficient." What he had realized was that under extreme conditions atoms could literally be crushed out of existence so that electrons were no longer associated with any particular nucleus, and that this situation allowed for an enormous compression of matter.

Eddington immediately saw a way to independently confirm this unexpected result: by measuring the gravitational redshift of Sirius B. In his discussion of this subject he clearly regarded the gravitational redshift not as a theoretical prediction, subject to observational test, but rather as a convenient tool to demonstrate the small radius of Sirius B. Eddington's frame of mind is very much in keeping with his view that General Relativity was self-evidently correct and not urgently in need of additional experimental confirmation. Eddington (Figure 9.1), after all, was instrumental in verifying Einstein's General Theory of Relativity during the 1919 British eclipse exhibitions that he led.

Eddington focused on Sirius B as an obvious target since, at the time, it was the only one of the three known white dwarfs, and the only one whose orbit and mass were already well established. Knowledge of the orbit is essential since it is necessary to separate the effects of the gravitational redshift from the Doppler shift due to any actual motion of the white dwarf with respect to the earth. It is the existence of the companion, with a measurable motion, which allows a clean determination of the gravitational redshift. Eddington's initial prediction of a gravitational redshift, based on the fact that Sirius B had close to the mass of the sun and (he assumed) only 1/30th its radius, was a value of approximately 30 km/s.

The next step was to make the necessary observations, which Eddington was well aware would be a challenge, due to the proximity of Sirius A. The obvious choice of observer and observatory was Walter S. Adams (Figure 9.2) and the Mt. Wilson Observatory located in the San Gabriel Mountains overlooking Los Angeles. It was Adams after all who had succeeded in getting the first spectrum of Sirius B in 1914, which helped establish that Sirius B had a relatively high surface temperature, and proved beyond all doubt that it must be exceedingly small. In the decade since Adams first observed Sirius B with the 60-inch reflector, it had been joined on Mt. Wilson by an even larger instrument: the remarkable 100-inch Hooker reflector, built though the untiring efforts of George Hale. The 100-inch became operational in 1917 and for the next 31 years, until the opening of the 200-inch at Mt. Palomar, this telescope would remain astronomy's premier telescope. Among other things, it would provide

Figure 9.1. Albert Einstein and Arthur Eddington, in Eddington's garden at the Cambridge Observatory (1930).

Figure 9.2. Walter S. Adams (Yerkes Observatory photograph).

Edwin Hubble in 1929 with the observations he used to establish the immense size and the expansion of the universe. In January 1924, before he delivered his paper containing his estimates of the size and density of Sirius B, Eddington had already written to Adams and suggested measuring the gravitational redshift of Sirius B:

> "I have lately been wondering if you would find it possible with your great instrument to measure the radial velocity of the companion of Sirius—with a view to determining its density by means of the Einstein shift of the spectral lines (compared with Sirius). ... The nearest prediction with the available data is 28.5 km per sec.
>
> Of course this involves a density of about 100,000 times that of water which is incredible. But I have recently been entertaining the wild idea that it may just be possible. Owing to the high ionization in the stars the atoms are almost certainly stripped down to the K-ring and I really do not see what is to prevent them packing close enough to give this density—if the pressure is high enough as it would automatically become.
>
> I should scarcely venture to suggest your following this wild idea if I did not regard a negative result as also of interest ... a very definite challenge to thermodynamicists ..."

Adams replied to Eddington's letter on February 12 mentioning that he had realized the Einstein shift, "should be large but [that he] had not attempted to compute it." He also revealed that he had spectroscopic plates of Sirius B, which he had considered measuring, but had always "laid them aside" due to the problems of scattered light from Sirius A. He promised Eddington he would make an attempt to measure the plates. The results were conveyed to Eddington in a letter dated March 3, stating "We have obtained some results on the companion of Sirius which I believe will interest you." Adams was able to measure four lines, including two due to hydrogen, and provided several preliminary estimates of the velocity shift. A simple mean gave +20 km/s, while giving double weight to the hydrogen beta line yielded +23 km/s. Stressing the difficulty of the measurements due to the "super-position of the spectra of Sirius and the companion," Adams suggested a correction factor of 1.5 appeared reasonable, giving, "... a result of not far from 30 km. This is in agreement with your prediction." Eddington wrote back on March 22 and took the opportunity to revise his earlier estimate of the redshift. He replied that he had originally assumed that the spectral type of Sirius B was "A", however, if he used a somewhat later spectral type of "F0", then the star would be cooler and consequently larger. This would result in a reduced redshift of 20–25 km/s as "the most likely value." Eddington's use of an "F0" spectral type for Sirius B cannot be explained. Eddington claimed to have found the "F0" classification in a "recent Mt. Wilson publication." There is no such publication. Moreover, if any subsequent classification work had been on Sirius B Adams would certainly have been involved. Yet the letters between Adams and Eddington indicate that Adams, the observer, made no effort to correct Eddington, the theoretician, on this point.

Adams replied on April 24, stating "I am glad that you took the results of Sirius as entirely provisional ..." and stated that he wished to confirm the observations and

"... would prefer to publish nothing until we can secure this." He closed his discussion of the topic by telling Eddington, "I shall look forward with greatest interest to reading your paper [Eddington's mass–luminosity paper] when it arrives. If by any chance the companion of Sirius should contribute a final bit of evidence to confirm your theory, it would be worth almost any trouble and difficulty."

The first opportunity for Adams to confirm his preliminary measurement came in the following year. Adams knew that his chief obstacle would be the contamination of his Sirius B spectra with light from Sirius A, and he took steps to minimize this. The telescope's secondary mirror was supported by four slender vanes orientated in the cardinal directions of the compass. In this orientation, the vanes produced four corresponding diffraction spikes, or rays, in the Sirius A image which fell close to the image of Sirius B. To solve this problem Adams masked the telescope aperture so that the vanes were covered and the diffraction spikes pointed away from Sirius B. The penalty for using the mask was the immediate loss of one-half the light collected by the primary mirror, but this loss was more than compensated for by the larger light-collecting capacity of the new telescope. The mask made the observation possible, but did little to mitigate the strong circular halo of scattered light from Sirius A which was also present. Adams tried to minimize this problem by selecting only those observations made under the best seeing conditions, when the scattering was least. He used the telescope's 36-inch spectrograph, which was located at the Cassegrain focus where the secondary mirror produces images through the central hole in the primary mirror. The spectrograph used a single prism to produce spectra that were recorded on 17-inch long photographic plates. A typical exposure required just 10 minutes.

Out of a number of observations obtained that season, Adams selected four plates that seemed to exhibit the least contamination from Sirius A. The Doppler shifts of the spectral lines seen in Sirius B were measured by his assistant, Miss Webb, in whom he placed great faith. Two different methods were used to estimate the relative displacement in the spectral lines from Sirius B with respect to the spectral lines in the scattered light from Sirius A. Adams recorded the measurements for the Balmer lines due to hydrogen, as well as for other lines due to such elements as magnesium, iron, and titanium. As will become apparent later, the presence of non-hydrogen lines in Sirius B came to have great significance. His various measurements differed appreciably, and he used a weighting scheme to arrive at a single average result. He also applied his best estimate for correcting the results for the scattered light from Sirius A. After finally accounting for the expected Doppler shift due to the orbital motion of Sirius B, Adams came to a final result of 21 km/s: in apparent agreement with Einstein's theory and almost precisely the revised value calculated by Eddington in March of 1924. Adams, well aware of the significance of his result, published it in July of 1925 in the prestigious *Proceedings of the National Academy of Science* in a paper entitled *The Relativity Displacement of the Spectral Lines in the Companion of Sirius*. So that his measurements not escape attention in Europe, and with the obvious approval of Eddington, the paper was also reprinted in the November issue of the *Monthly Notices of the Royal Astronomical Society*. It was Adams who highlighted this measurement not primarily as a confirmation of Eddington's prediction of a small radius for Sirius B, but as a third, independent observational verification of General Relativity. This is not surprising since a major effort to measure the solar gravitational redshift had been underway for a number of years at Mt. Wilson by the solar astronomer Charles St. John, without any clear-cut success. The Sirius B measurement was widely accepted at the time and has gone down in popular history as the third and final of the three "classical" verifications of Einstein's Theory of General Relativity.

Eddington quickly saw the public relations advantage of casting the Sirius B observations as a verification of General Relativity, as well as confirmation of his theoretical prediction of the radius of Sirius B. In his 1926 classic book, *The Internal Constitution of the Stars*, he described Adams' results in considerable detail, obviously satisfied that both the validity of General Relativity and the reality of the small size and immense density of white dwarfs had been demonstrated with a single elegant observation.

It was later pointed out that Adams' value for the redshift required a small change due to a mistake in his calculation of the orbital velocity of Sirius B. The basic result, however, was soon confirmed by Joseph H. Moore in 1928, using the 36-inch refractor at the Lick Observatory. Moore also obtained an identical value of 21 km/s, but admitted that this was largely a coincidence since the result contained an uncertainty which he estimated at ±5 km/s. There would be no further attempts to improve on the observations of Adams and Moore for Sirius B for another 40 years.

When Eddington initially calculated the gravitational redshift for Sirius B, he had a reasonably good grasp of the mass of the star, 0.85 solar masses, from its orbital motion. What he lacked was a comparable estimate of the stellar radius. He knew from its spectral type it must be small, but how small? He used an assumed temperature, based on his "F0" spectral type, of 8000 K to estimate a radius of 19,600 km. We now know (see Chapter 12) that his temperature was a gross underestimate, the actual value is at least three times higher and consequently the stellar radius is much smaller: all of which leads to a gravitational redshift that is *four times* that estimated by Eddington and measured by Adams. This particular difficulty was not initially apparent. However, there were early indications that the gravitational redshift for Sirius B was difficult to reconcile with the theoretical radius of the star. In contrast to the non-relativistic models favored by Eddington, Chandrasekhar's theory of white dwarfs gave much smaller stellar radii and thus larger redshifts. That theory, however, contained an adjustable parameter, the mean molecular mass of the matter in the degenerate core. Chandra, following the convention at the time, used a mean molecular mass that assumed a significant fraction of hydrogen, which lowers the mean molecular mass. This made sense, because it was discovered a few years earlier that hydrogen was the most abundant element in stars and, after all, Adams and Moore had both observed strong hydrogen lines in the spectrum of Sirius B.

By the late 1930s, questions began to be raised about both the radii of white dwarfs in general, and Sirius B in particular. Additionally, there were questions about the consequences of significant amounts of hydrogen (protons), at the temperature and densities that must characterize the central regions of a white dwarf. The issue of white dwarf radii was pressed by Gerard Kuiper, an astronomer at Yerkes Observatory. He realized that if he could accurately determine the temperatures for white dwarfs of known distance he could effectively measure their radii. In the mid-1930s Kuiper intensively studied the few then recognized white dwarfs, as well as going on to discover many more himself. Abandoning the early crude estimates of temperature, based on the superficial similarities of white dwarf colors with the colors of normal stars, Kuiper methodically studied both the precise distribution of light with wavelength as well as investigating the details of the shape of the broad hydrogen lines in their spectra. His use of the latter method was well in advance of his time and allowed him to make the first credible estimates of both the temperature and the pressure in white dwarf atmospheres. He showed that in general white dwarfs were actually hotter than early estimates. This meant smaller radii, which favored Chandrasekhar's theory. For one star, 40 Eridani B, Kuiper obtained a temperature that was not too far from the modern value. By this time, a mass estimate for 40 Eridani B had been determined from its orbit. Kuiper was therefore able to compare his result with Chandresekhar's mass–radius relation for white dwarfs. The agreement was remarkably good, provided that the degenerate core of 40 Eridani B was composed of material with a mean molecular mass of 2 atomic mass units: that is, it contained little or no hydrogen.

Sirius B was the only other white dwarf for which there was any hope at the time of determining both a mass and a radius. Kuiper never observed Sirius B himself. However, while he was working at Lick Observatory, Kuiper did examine the spectra that J. H. Moore had obtained in 1928, which had confirmed Adam's gravitational redshift. Kuiper was clearly perplexed by the appearance of the spectra. They looked nothing like those of the other white dwarfs he had studied: the hydrogen Balmer lines were far too narrow. Nevertheless, he made a temperature estimate of 9000 to 10,000 K, somewhat hotter than Eddington's estimate. When he plotted the resulting radius on Chandrasekhar's mass–radius relation, Sirius B fell well off the curve that nicely fit 40 Eridani B. The discrepancy implied that Sirius B had a larger than expected radius and thus it had to be composed of material with a lower mean molecular weight. This meant that the interior of the star must contain a significant admixture of hydrogen. Even with this larger radius, Kuiper was unable to reconcile the result with the gravitational redshift measured by Adams and Moore. Any reasonable mass and radius for Sirius B would he felt result in a gravitational redshift much larger than that found by Adams. In desperation, Kuiper asked a critical question: What would happen if he totally disregarded his estimate of the temperature and also assumed there was no hydrogen in the degenerate core? The answer he came to was that the temperature would be 25,000 K and the redshift would be +76.6 km/s. Both estimates he believed, given the measured gravitational redshift, were equally impossible—ironically modern measurements (Chapter 12) show that he was almost spot on with regard to both the temperature and redshift. As we will shortly see, Kuiper was not the only person to be misled by the Adams and Moore spectra of Sirius B.

In 1939, when Hans Bethe at Cornell University finally worked out the nuclear reactions and reaction rates for the net conversion of hydrogen to helium in stellar cores, it then became reasonable to suspect that no hydrogen could remain in the core of a white dwarf. The problem of determining the temperature of the degenerate matter in white dwarfs was taken up in 1940 by Bethe's graduate student, Robert

Marshak (1916–1992). Although temperature is not critical to the degenerate electrons, it is significant for the nuclei in the stellar core. What Marshak found was that the temperature in the cores of both Sirius B and 40 Eridani B was high enough that hydrogen would readily react to form helium and, moreover, if even a small amount of hydrogen remained, the resulting nuclear burning would produce far more luminosity than was observed. The likely composition of both stars was predominantly carbon and oxygen nuclei, in which case the mean molecular mass must be close to 2 atomic mass units. An obvious implication of this finding was that Sirius B had to be much smaller than Kuiper estimated, and the Adam's redshift of Sirius B was, in all likelihood, seriously at odds with this smaller radius. Marshak used Kuiper's best observational estimates for Sirius B, and even quoted Kuiper's precociously insightful estimates of temperature of 25,000 K and a gravitational redshift of 80 km/s for the star. Unfortunately, there was little hope at the time of repeating Adams' measurement, since Sirius B was at the time hidden in the glare of Sirius A.

Marshak also helped clarify some of the internal physics that must pertain to Chandrasekhar's white dwarfs by considering the thermal, radiative, and ionization conditions inside these stars. Taking as examples Sirius B and 40 Eridani B, he calculated the internal thermal energy and estimated the minimum cooling lifetimes by dividing total thermal energy by the current luminosities of these stars. His respective ages were 300 million and 500 million years, which he noted were in agreement with the known space density of white dwarfs as well as the then estimated age of the universe of one billion years (the universe is actually 13.7 times older than this!). This was one of the first attempts to determine the ages of white dwarfs. Marshak also identified some of the processes that must occur inside a white dwarf. The luminosity of white dwarfs was supplied almost entirely by the huge reservoir of thermal energy residing in the degenerate core, which was slowly being released by thermal conduction. In Sirius B the core was essentially isothermal, at a temperature of 15 million degrees Kelvin, since the free electrons in the core made it an excellent heat conductor. Beyond the degenerate core, energy would be transported to the surface by radiation and convection and would finally emerge from a thin atmosphere dominated by hydrogen.

In the early 1960s, Sirius B was again approaching its maximum separation from Sirius A. The intervening four decades had witnessed a dramatic improvement in the ability of astronomers to interpret the spectra of stars and of white dwarfs in particular. New, very comprehensive calculations of the white dwarf mass–radius relation were now available and they confirmed essential elements of Chandrasekhar's early models. In addition, new and improved instruments were in place and the time had come to repeat Adams' observation of Sirius B. This time (1961–1962) the observers were Drs. J. Beverly Oke and Jesse Greenstein at the California Institute of Technology and the instrument was the Palomar 200-inch telescope. As with the 1924 observations, an aperture mask was required, but the observers were able to take advantage of the fact that the 200-inch mirror had just been given a fresh coat of aluminum and was in prime condition to deliver images of Sirius A with a relatively low degree of scattered light. Again, photographic plates were used to record the spectra. This time, however, obtaining a measurement of the gravitational redshift

Figure 9.3. The relative sizes of Sirius B and the Earth (Space Telescope Science Institute).

was not the only goal. The same spectra which recorded the redshift also provided good measurements of the shape, or profile as a function of wavelength, of the hydrogen lines in the spectrum of Sirius B. The new spectra showed no trace of the magnesium, iron, and titanium lines seen by Adams, only broad deep hydrogen lines, in contrast to the narrow shallow lines present in the Adams and Moore spectra. The analysis of these spectra was conducted by a young graduate student at Caltech, Harry Shipman, and published in 1971. From the observed shape of these lines, together with detailed theoretical calculations of the expected line shapes, the observers also were able to make a revised estimate of the surface temperature of Sirius B ($32{,}000 \pm 1000$ K) and a gravitational acceleration at the surface of Sirius B ($\log g = 8.65$). The final result for the gravitational redshift was $+89 \pm 16$ km/s; a result considerably at odds with those obtained by Adams and Moore but as Greenstein and colleagues state, precisely the new expected value.

It is now also finally clear that the estimates that Eddington had used for the temperature of Sirius B were much too low and this led him to assume a much larger radius for Sirius B than is actually the case. The currently accepted value is 5840 km, or about 92% the radius of the Earth (Figure 9.3, see also color section). This oversight largely explains Eddington's erroneous calculations but provides no insight into how Adams, followed by Moore, measured a redshift that was a factor of four too low.

Clearly, what happened was both observations were contaminated by scattered light from Sirius A. This is evident from the spectral lines due to the heavy elements of magnesium, iron, and titanium measured by Adams, which actually belong to Sirius A. These elements are now known to be absent from the uncontaminated spectrum of Sirius B. Nevertheless, Adams also measured the hydrogen lines in the Sirius B spectrum and still got an incorrect result. This has led some historians of science to question the Adams' measurement of the gravitational redshift, citing his result as a prime example of an observer knowing "the correct answer" in advance and then proceeding to find what was sought. The chief critic is Norriss Hetherington of the University of California at Berkeley, who published several articles questioning the competence, objectivity, and professionalism of observers during the early years at Mt. Wilson, in particular Adams.

Walter S. Adams was a highly respected observer at the Mt. Wilson Observatory in the period before and after the First World War. At the time of the redshift measurements, he was director of the observatory and in 1928 he was awarded the prestigious Bruce Medal of the Astronomical Society of the Pacific. He was one of the principal observers using the 100-inch Hooker telescope, primarily for spectroscopy. In the pages of the 1980 *Quarterly Journal of the Royal Astronomical Society*, Hetherington employs terms such as "alleged", "purported", and "aberrations" to describe Adams' redshift measurements and argues that Adams fell victim to the scientific vice of abandoning objectivity in order to find exactly what he set out to find in the data. Hetherington goes on to scourge Greenstein and company with quotations from Shakespeare's *Richard II* for corrupting the integrity of the scientific process by glossing over the clear disagreement of their 1971 results with Adams' 1924 measurements. Since Greenstein and his coauthors were also part of the Mt. Wilson establishment, Hetherington attributes the motivation for this lack of curiosity of how Adams could have obtained such an obviously incorrect result to institutional fealty.

In 1985, Greenstein, Oke, and Shipman responded to Hetherington's charges in the pages of the same journal where he had leveled them five years earlier. They defended their integrity, as well as that of Adams, by pointing to the extreme difficulty of the measurements that Adams made. His instruments, while the best in the world at the time, were crude by contemporary standards. Greenstein and Oke went on to describe their efforts to reexamine the sixty-year-old photographic plates used by Adams and to comb the observatory observing logs of the period in an effort to reconstruct his observing procedures. In the end they identified several plausible factors which contributed to Adams' results. The first and most important was the halo of scattered light from Sirius A. Adams recognized this problem and attempted to correct for it, but he clearly underestimated the magnitude of the effect. Moreover, he had no reason to believe that the lines of magnesium, iron, and titanium were not part of the spectrum of a white dwarf such as Sirius B. Other problems were traced to the spectrometer and the manner in which its slit was illuminated.

In a second following paper, Professor François Wesemael of the University of Montreal undertook a theoretical calculation of the effect of scattered light from Sirius A on the measurement of the hydrogen lines in Sirius B. While it is not possible to fully reconstruct Adams' observations, it is possible to estimate the approximate levels to which the observations of Sirius B are contaminated on the original spectroscopic plates. Dr. Wesemael used accurate theoretical representations of the relative spectra of each star and numerically combined them in proportions which represented the expected range of contamination. He was able to demonstrate that as the relative level of contamination increases, the "measured" redshift becomes smaller. Indeed, a contamination level of 50% to 75% would easily explain Adams' results. Adams was clearly wrong, but it appears there are no grounds to question his integrity with regard to the redshift of Sirius B.

There remain some ironic footnotes to the saga of Sirius B and the gravitational redshift. Observational confirmation of the gravitational redshift is no longer sought in the stars. Its reality was finally established to everyone's satisfaction here on earth, in a physics laboratory on the Harvard University campus. In 1960 Professors Pound and Rebka devised a precise and elegant demonstration of the gravitational redshift by measuring the redshift of gamma rays emitted and absorbed by ^{57}Fe nuclei (the Mossbauer Effect) located at the top and bottom of a 22-meter high tower. Although the effect on earth is some 1000 times smaller than on Sirius B, Pound and Rebka were able to measure the gravitational redshift of the gamma rays to within 1% of the predicted value. An even more precise test of the gravitational redshift was performed in June of 1976 when a sounding rocket was launched to an altitude of 10,000 km above the earth's surface over the Atlantic. On board was a hydrogen maser (the microwave equivalent of a laser and a very accurate clock) oscillating at a precise frequency. As the rocket climbed towards the peak of its trajectory and then fell back to earth the frequency of the maser responded precisely to the changing gravitational field of the earth as predicted by General Relativity to better than 7 parts in 10^5. More fundamentally, it is now widely appreciated that the gravitational redshift is not actually a crucial test of the Theory of General Relativity itself, but rather a direct consequence of the more primitive concept of the equivalence between gravitational and inertial mass. Astronomical interest in the gravitational redshift now focuses on its use as a tool for measuring the masses and radii of compact objects such as white dwarfs, neutron stars, and black holes; an objective that Eddington would have readily approved. In Chapter 12 we will consider just such a space age determination of the redshift of Sirius B and how it can be used to help improve our knowledge of the mass and radius of this star.

Part Four

A Controversial and Occult Sirius

Sirius, perhaps more than any other star, has inspired a host of controversies. Chief among these is the notion, prompted by Ptolemy's famous comment, that in ancient times Sirius was a fiery red star and that it changed its appearance some time between the 2nd century and the 9th century AD. A more modern controversy is the idea that the Sirius System harbors a faint third companion. Sightings of such a third component have been reported from time to time, and it is held to be responsible for small perturbations in the orbit of Sirius that have been found by some investigators since the 1930s. These aspects of Sirius are amenable to scientific investigation. However, there are also a number of persistent beliefs that have become part of modern popular culture that have taken on a life of their own. These include the mythology of the Dogon tribe in the African nation of Mali, a best-selling book postulating ancient aquatic visitors from Sirius, and a cult whose beliefs in Sirius helped propel them to oblivion.

10

A Red Sirius

"The star in the mouth, the brightest, which is called 'the Dog' and reddish"
—Claudius Ptolemy, *Almagest*, Book VIII, 150 AD

One of the most contentious and long-running mysteries regarding Sirius originated in the 2nd century AD with what appears to be a casual comment made by the Alexandrine astronomer/astrologer Claudius Ptolemy (Chapter 3). Books VII and VIII of Ptolemy's *Almagest* contain one of the earliest and most famous of the ancient star catalogs, in which Ptolemy lists the positions and brightness of some 1022 stars. He comments on the color of only six of these stars—Betelgeuse, Aldebaran, Pollux, Arcturus, Antares, and Sirius—and assigns the color red to each. In particular, for Sirius in the constellation Canis Major, he states its location, on the dog's mouth, as well as its relative brightness and color: bright and red.

To modern eyes Ptolemy's descriptions of color are accurate for the six stars, with the singular exception of Sirius, which is clearly not red but bluish white. Was Ptolemy mistaken concerning the red color of Sirius, or was this merely a transcription or translation error that occurred somewhere in the generations of copied manuscripts from the original Greek to Arabic and to Latin? More significantly, as some have suggested, perhaps Ptolemy was correct and Sirius did appear red in color two thousand years ago, and sometime during the intervening two millennia it became white. If so, then perhaps Ptolemy has left us with a valuable historical clue to the evolution of the stars. Alternatively, there perhaps exist other explanations more rooted in human cultural history. It is this brief passing reference to the color of Sirius by Ptolemy and its implications for the evolution and nature of stars that launched a 250-year-old controversy.

The strange case of the "Red Sirius" begins in earnest in England in 1760 with the local squire of Lyndon-Hall in Rutland, Thomas Barker, who was an amateur astronomer and meteorologist. In January of that year a paper by Barker was read at a meeting of the Royal Society in London. The paper inquired into the brightnesses and colors of the stars and planets as perceived by the ancients and whether any of these bodies had undergone observable changes since that time. Barker noted that since antiquity several naked eye stars had perceptibly faded and some had disappeared altogether, while a few new stars had been discovered. He then recalled the two transient stars, recorded nearly a century earlier by Tycho Brahe and Johannes Kepler, that suddenly brightened and then changed color as they slowly faded (see Chapter 3). But had any star ever been observed to entirely change its color? In general he concluded:

> "But, I think, no one has yet remarked, that any lasting star was of different colour in different ages."

Although the colors of the stars and planets had remained largely unchanged, Barker called attention to the prominent exception of Sirius. One of the tenets of astronomy inherited from the ancients, and Aristotle in particular, was that the heavens were immutable and not subject to change. However, stars were known to change their brightness and the proper motions of stars were, by Barker's time, an established fact. Therefore, might their colors also change? Barker then proceeded to produce Greek and Roman authors, including Aratus, Seneca, Horace and Cicero, all of who appear to have regarded Sirius as red in antiquity. Barker's evidence was not widely accepted until Sir John Herschel (son of Sir William Herschel) commented favorably on the idea in 1839. While in South Africa, Herschel had witnessed the famous "Great Eruption" of the star Eta Carinae, which began in 1837 and for a brief period in 1843 it had shone at −1 magnitude, nearly as bright as Sirius.

The principal modern proponent of a historically Red Sirius was the remarkable T. J. J. See, whose astronomical career spanned seven decades. Thomas Jefferson Jackson See was born into a prosperous Missouri farm family in 1866. A tall and imposing young man (Figure 10.1), See attended the University of Missouri, where he excelled in virtually all subjects, particularly the sciences. During the commencement of June 1889, young See was the class valedictorian, receiving degrees in the arts, sciences, and law as well as honors in several fields. In addition to his academic achievements, See's university years revealed several personal traits which were to cause him a great deal of trouble throughout his later astronomical career. See apparently led a student faction that was instrumental in ousting the university president. He was also involved in a murky affair resulting in an accusation of plagiarism, after he won a gold medal in oratory. Although he was exonerated, the aftermath left such bitterness that the annual oratory contest was abolished. It was also at university that it became clear that See harbored a burning ambition and a hunger for acclaim—which he pursued with a single-minded intensity throughout his life. At the University of Missouri, See studied under W. B. Smith, a mathematics instructor whose primary interest was biblical criticism. It is from Smith that See obtained his lifelong interest in classical studies, which he freely mixed with astronomy on more than one occasion.

Figure 10.1. Thomas Jefferson Jackson See (frontispiece, *Brief Biography and Popular Account of the Unparalleled Discoveries of T. J. J. See*, W. L. Webb, 1913).

After graduation, See pursued graduate studies in Germany at the University of Berlin where he acquired an expertise in binary stars. He graduated *magna cum laude* with his doctorate in 1892 and immediately secured a position at the newly established University of Chicago under George Ellery Hale, who would later go on to promote the establishment of the large reflecting telescopes at the Mt. Wilson and Mt. Palomar Observatories. At the time, however, Hale was busy building the Yerkes Observatory, centered on the 40-inch refracting telescope with its famous Alvan Clark objective lens. It was in Chicago that the trouble started. See worked on an ambitious program, led by Sherwood Burnham, to improve and recompute a number of binary star orbits. See had already established a solid reputation in American astronomy, frequently publishing his binary star work in both American and European journals. At this time his publishing activity also extended to the popular press in such venues as *Harper's Weekly*, *Atlantic Monthly*, and *Popular Astronomy*, where he became a frequent contributor on a variety of subjects. In 1892 See first began publishing his investigations of the Red Sirius question from the vantage point of classical literature.

Difficulties emerged when See's ambitions at Chicago began to conflict with those of Hale. The newly established astronomy department and Yerkes Observatory were having financial problems and See proposed a fiscal plan to the University president to cut operational expenses. Perhaps prompted by this move, a dispute developed between Hale and See regarding the funding for the publication of a book written by See, which consisted primarily of his previously published binary star orbits. In addition, See sought promotion to associate professor, a rank equivalent to Hale. He took his case directly to the University president but received no support. In 1896 See and the University of Chicago parted company on bad terms.

See had succeeded in establishing himself as a well-known observer, and on this basis he was hired by Percival Lowell to help establish Lowell's new observatory in Flagstaff, Arizona. It was at Lowell Observatory, in the spring of 1896, using the 24-inch Alvan Clark lens, that See made his premature announcement of the recovery of the companion of Sirius (Chapter 6). Again, See soon clashed with his colleagues. When not observing Mars, Lowell resided primarily in Boston and A. E. Douglass was the effective day-to-day director of the observatory. In 1897, while Douglass was ill and recuperating in San Diego, See was placed in temporary charge of Lowell Observatory. Upon his return, Douglass found that most of the observatory staff had quit, unable to put up with See. Douglass wrote a devastatingly frank letter to one of the observatory's Boston trustees, detailing See's many faults and including reference to a confidential physician's evaluation which contained a diagnosis of mental aberrations and personal vices including "cowardice", "moral obliquity", and "loss of sense of right and wrong as well as mental peculiarities". To this, Douglass added his own graphic account of his personal loathing of the "vile" See, even warning of the danger of fistfights between See and the staff, if he remained. Before the end of 1897 See was fired from Lowell Observatory.

Again See landed on his feet, this time through his old Missouri political connections including the speaker of the U.S. House of Representatives. He managed to secure an appointment in 1899 from President William McKinley to the staff of the U.S. Naval Observatory, as professor of mathematics. At the Naval Observatory, See continued his career as an observer with access to its superb 26-inch Clark refractor. In 1902, having suffered a nervous breakdown and after a medical leave of absence, See was transferred to the U.S. Naval Academy at Annapolis as a mathematics instructor. When his health failed to improve, he was finally transferred to an obscure U.S. Naval time-keeping station at Mare Island near Vallejo, California. He was only 37 years old and no longer had access to telescopes, other than a small 5-inch Clark refractor, of which he made little apparent use. Although this effective exile ended his observing career, it launched him on another equally controversial career as both a prolific writer of dense theoretical works on physics and astronomy, and as a widely read popularizer of science. His official duties were apparently not demanding, since they left him with ample time to produce voluminous books and articles outlining his own theories of the universe. In 1913 See even underwrote the publication of his own biography, *A Brief Biography and Popular Account of the Unparalleled Discoveries of T. J. J. See*, by W. L. Webb.

See never avoided controversy; indeed he seemed to seek it out. In one peculiar episode in 1895 See published a paper analyzing the motion of the binary star 70 Ophiuchi and reporting the discovery of a dark third component of the system.

In effect, he was claiming to have accomplished what Bessel had done with Sirius: predict the existence of an unseen star from the motion of the visible components. Shortly after the paper was published, F. R. Moulton, a former graduate student who had worked under See at Chicago, published an analysis of this paper showing that See's unseen companion could not possibly exist, since the claimed orbit was unstable. This episode coupled with See's defensive reply earned him a virtual ban on publishing in the *Astronomical Journal.* He did continue to publish in Europe, but his ideas were not widely accepted by his colleagues. In spite of this setback he remained an effective public spokesman for his own unorthodox ideas, which received wide circulation in the press and popular publications of the day. In 1909 he developed his own comprehensive theory of the formation of the solar system that he called the "capture theory". See laid the details of this theory out in the second volume of his *Researches on the Evolution of Stellar Systems*, which he had published, largely at his own expense in 1910. In 1917 he extended his theoretical domain into the realm of fundamental physics with his theory of waves in the aether that he claimed provided explanations of gravitation and electromagnetic phenomena. See rejected Einstein's General Theory of Relativity and, in particular, the result predicting the gravitational deflection of starlight by the sun, and he publicly accused Einstein of having made fundamental errors. Later he also disputed Edwin Hubble's interpretation of the redshifts of external galaxies as being due to an expanding universe. See contended that his aether wave theory could account for the redshifts in terms of the energy lost by light in collisions with interstellar dust.

T. J. J. See died in 1962, at the age of 96, having outlived most of his adversaries, and all but forgotten by the astronomical community. He remained firm to the end in his belief that he had solved many of the major problems of physics and astronomy. His ideas, however, still appear to hold some fascination for those advocating "vortex" theories of the universe. It would be unfair and misleading to completely dismiss See as a mere crank or intellectual charlatan. He was a classically trained visual astronomer, whose work on double stars was generally respected. He never managed to gain much appreciation for the new physics and astrophysics that developed early during the 20th century and he tenaciously clung to the classical physics he learned as a student. It was, however, his ambition, enormous ego, limitless capacity for self-promotion, as well as combative personality that led to most of his difficulties.

While a graduate student in Berlin, See began a long-running investigation of the Red Sirius issue. Indeed in one of his publications he relates a voyage undertaken in March 1891 to the city of Alexandria, in search of the observatories where Ptolemy and Hipparchus had once observed the skies, recalling it as "... one of the greatest inspirations of my student days." See's interest in historical accounts of a Red Sirius resulted in a series of scholarly papers published in 1892, followed by a less academic, abbreviated version in *Popular Astronomy* in 1895, and culminating with a definitive summary of his studies in 1926.

In this latter work See boldly states his case and defines his methods. He begins with an appeal to contemporary ignorance, arguing that much was still to be learned concerning stellar evolution: therefore in the absence of a comprehensive theory, anything was possible. In this he was partly correct, since at the time such organizing principles as the Hertzsprung–Russell diagram were in the process of being introduced and the understanding of the nature of nuclear reactions, which control the evolution of the stars, was nearly three decades in the future. He goes on to argue that if we can't precisely know the future, we can at least clearly discern the past, stating "... yet in the matter of the color of Sirius I insisted from the outset that the problem was purely one of historical fact." He lays out his strategy: "... Moreover, under valid methods of procedure, historical questions, in general, are settled by the evidence of the ancients,—the only persons who could possibly bear witness to the truth of phenomena transpiring so long ago." He acknowledges that there are skeptics, but declares that they have fallen into error by habitually citing the flawed scholarship of earlier workers. For example, of the Danish astronomer G. Schjellerup who in 1874 had cast doubt on some of Thomas Baker's original sources, See said "And even in later times Schjellerup's errors have had considerable circulation." See also chides the preeminent 19th century American astronomer Simon Newcomb who dismissed the color change of Sirius in his 1908 textbook, as being, "... not inclined to admit any evidence of antiquity as sufficiently strong to establish a change in the color of Sirius, it can hardly be said that he looked at the question on its merits, but rather arbitrarily pronounced an adverse decision." See is also critical of the unconverted: "If a modern astronomer, on general grounds of improbability, pronounces their [*the ancients*] statements inadmissible, the step implies an assumed understanding of the laws of Nature, and thus there will be no way to convince so presuming an unbeliever; for he admits a procedure, entirely unwarranted by the recognized cannons of historical criticism." Finally, See assures the reader that he is not alone in this search for historical truth and that "... And a good many careful writers in Germany, Sweden, Russia, America and other countries have come to a similar conclusion."

Having laid bare the faulty scholarship of his critics, he chastened those who failed to properly consider his own work. See marshals his evidence in a form more resembling a legal brief, than a scientific paper. Hosts of Greek and Latin writers and poets are summoned to testify to the "color" of Sirius in antiquity. Exhibit A is the single line concerning the color of Sirius from Ptolemy in the *Almagest*. He carefully establishes Ptolemy's credentials as the foremost ancient astronomer and reviews the chain of scholarly custody of Ptolemy's text from the Greeks through the "Saracens" and the Persians to Medieval Europe. He notes that there is no mention of a Red Sirius in Islamic works, beginning in the 9th century and uses this to establish the time frame for Sirius' color change from red to white. He bolsters his case with supporting testimony from the now familiar sources such as Aratus, whose 3rd century Latin poem *Phaenomena* describes Sirius as "ruddy". He then moves on to 1st century Rome and the orator Cicero and Germanicus (the popular general and son of the emperor Tiberius). Although these authors are not astronomers and are only translating or commenting on Aratus' poem, See assures us that they are careful sober witnesses who must have been familiar with the appearance of the heavens. Next the Roman poets Horace, Virgil, and Manilius take the stand and describe Sirius in colorful, but ambiguous terms. Perhaps the most explicit and credible testimony comes from the 1st century writer and naturalist Pliny who describes Sirius as "fiery" (*igneus*) and the philosopher Seneca who describes the color of Sirius as a deeper red than that of Mars ("... Caniculae rubor, Martis remissior ..."). A number of other authors from Homer to Hesiod are questioned, but their descriptions of Sirius are far from conclusive. See seems to have exhaustively and carefully surveyed classical literature for references describing Sirius and gives the confident appearance of being at ease with the subtleties of the original Greek and Latin texts.

There is, however, no effective cross-examination of these sources, nor is any testimony from Greek and Roman sources ever presented that would contradict these quotes and descriptions. For example, the Italian astronomer Giovanni Schiaparelli had noted prior to See's efforts that the 1st century AD Roman poet Marcus Manilius referred to Sirius as "sea-blue". Later others cited the 4th century Roman poet Rufius Festus Avienus' use of the same term to describe Sirius. See appears to be meticulous in his elaboration and interpretation of the meanings of the ancient passages. After reading See, one is left with the inescapable conclusion that the Greco-Roman world was unanimous in its judgment that Sirius was once a red star, and not at all the intense blue–white star we see today. See sums up his exhaustive literary investigation of the ancient color of Sirius as follows: "And thus if I have restored the authority of the greatest of Greek astronomers, as to the aspect of the brightest of the fixed stars, at the most flourishing period of the School of Alexandria, and thus rescued a priceless and everlasting record from threatened oblivion, the object of the historical researches entered upon at Berlin 35 years ago will have been attained."

See's last major contribution to the cause of a Red Sirius was in 1926 when he published an enhanced version of his arguments in the German journal *Astronomische Nachrichten*. After this, interest in the subject faded among the astronomical community, as understanding of the subjects of stellar spectra and stellar evolution advanced. There seemed little room for stars such as Sirius to abruptly change colors, after all this would imply that the surface temperature of Sirius would have had to increase from a few thousand degrees Kelvin to nearly 10,000 Kelvin in a few centuries.

The most recent reemergence of the "Red Sirius" controversy began in the "Correspondence" pages of the British journal *Nature*. In 1983 a brief letter appeared from two Canadian readers inquiring about Homer's frequent use of the descriptive term "wine dark sea" or "wine coloured sea" and whether this might have come about from the Greek practice of diluting wine with water, so that if the local water were sufficiently alkaline, then it may have raised the pH of the wine enough to change its color. Over the next several years this rather arcane academic discussion prompted a host of follow-up letters offering other explanations of what Homer's poetry really meant or what he was trying to describe. Some contributors suggested the phrase referred to the appearance of the sea at sunset under dusty conditions or the color of wine poured into ceramic bowls or bronze goblets. Finally, in 1984 two astronomers wrote independently to suggest that the ongoing discussion reminded them of the red color of Sirius, as described by Ptolemy. At this point, the long-running discussion shifted abruptly from the color of wine and the Mediterranean during the Bronze Age, to the way various ancient cultures had described Sirius. One Cambridge reader noted that in Icelandic mythology Sirius was known as "Loki's Brand", due in part to its flaming color. R. H. van Gent, an astronomer in the Netherlands, contributed a reference to an Assyrian text from 1070 BC mentioning the color of Sirius as "red as molten copper". Dr. van Gent, however, noted that this description referred to the rising of Sirius near the horizon and cited other Roman and Chinese sources from 90 BC to 625 AD, which referred to Sirius as white or bluish. This was followed by Schlosser and Bergmann, two German astronomers, who reported an early-medieval manuscript that appears to describe Sirius as red. The late 6th century Latin manuscript entitled *De cursu stellarum ratio* was from the Lombard region of France and due to St. Gregory of Tours. It contained instructions for determining the hour of nighttime monastic prayers from the positions of the constellations, and referred to Sirius as "Rubeola" or reddish. Schlosser and Bergmann hypothesize that it was Sirius B, in its red giant phase that was responsible for its red colors. They went on to estimate that if this were the case Sirius would have not only appeared red but also much brighter at a V magnitude of −4, approximately equal to Venus at its brightest. However, van Gent and Stephen McClusky disputed this interpretation, noting that it is much more likely that St. Gregory was describing the red star Arcturus and not Sirius.

The discussion in *Nature* next turned to the historical record from the Orient. In 1986 Tong Tang of the Hong Kong Baptist College cited a passage from Han dynasty historical records of two thousand years ago, describing the color of several stars: "The white is like Sirius, the red is like Antares and the yellow like Betelgeuse ..." This would seem to provide an airtight alibi for Sirius against the charge of "redness". However, four years later two French astronomers, Cécile Gry and Jean-Marc Bonnet-Bidaud, using the same document as Tang, found the following passage: "At East there is big star called Wolf [Sirius], Wolf horn changes color, many thieves robbers." Perhaps, they argued, Sirius had changed color after all. Tang replied that if the passage were read in its proper astrological context it means that if Sirius was observed to change color, then the occurrence of thieves and robbers was to be expected. Jiang Xiao-yuan of the Shanghai Observatory provides the most definitive discussion of the color of Sirius in ancient Chinese records. He published a study of the way various bright stars, and Sirius in particular, were referred to over time. He noted that according to Chinese astrological tradition, color was an important attribute of stars and planets and was therefore carefully noted. Observers would compare the colors of stars with particular standard stars, one of which was Sirius. Jiang Xiao-yuan found four separate records of the color of Sirius during the period 100 BC to 650 AD, all of which described stars as being "as white as Lang", Lang being the Chinese word for Wolf or Sirius. Thus, the evidence from ancient China seems to argue strongly against any change in the color of Sirius in the last two millennia.

One modern explanation for the possible reddish appearance of Sirius involves not the star itself, but interstellar dust. The reddening of starlight by grains of interstellar dust is a well-known and carefully studied astronomical phenomenon. Grains of interstellar dust—smaller than 0.1 micron and composed of carbon, silicon, and oxygen—are formed in the outer atmospheres of red giant stars and dispersed into interstellar space as these stars shed their outer envelopes. When the line of sight to a star contains enough interstellar grains they can perceptibly dim and redden starlight by preferentially scattering blue light. Over the last 25 years a number of astronomers have suggested that two thousand years ago Sirius was obscured by a small dense cloud of interstellar dust that caused it to appear red, but which has now moved out of the line of sight. Observational evidence of this cloud has been searched for in the vicinity of Sirius without success. Dr. Douglas C. B. Whittet at the Rensselaer Polytechnic Institute in New York, however, gives a strong theoretical argument against this explanation. Any cloud sufficiently dense to redden Sirius to the extent that it would resemble in color the red stars Arcturus or Betelgeuse, would also dim its light to the point it would appear as a mundane 3rd magnitude star. Moreover, Whittet notes that the human eye does not perceive color in stars this faint. Thus, a Sirius obscured by an interstellar dust cloud would be an unremarkable 3rd magnitude star and not be seen by the naked eye as red at all.

So why did Ptolemy and other ancients consistently refer to Sirius as red or reddish, when it is so clearly intensely white or bluish? There are several possible answers to this question. First, as pointed out by Dr. Roger C. Ceragioli (see Chapter 2), who has written extensively on the ancient references to a Red Sirius, many descriptions cited by See and others are concerned with the intense twinkling and diamond-like flashing of Sirius caused by turbulence in the earth's atmosphere. The important point is that only intrinsically white stars such as Sirius possess the full complement of wavelengths necessary to produce these multi-colored scintillations. Red stars lack the short wavelengths and therefore cannot appear to brilliantly flash in the same fashion and these would not have been remarked on by the ancients. Second, in many parts of the ancient Mediterranean world the "heavens" were the stage on which events of astrological and religious significance unfolded. Many primary heavenly bodies—such as the sun, the moon, and the planets—possessed associated attributes, such as the red color of Mars being connected with war and strife. In this way, the attribute and the heavenly body were both linked in the popular mind, and were reflected in the phrases used to describe these bodies. Thus, Sirius when it appeared out of the solar glare in late July was associated with the arrival of the oppressive heat of late summer that in turn was variously associated with droughts, searing heat, and flaming tongues of fire and fevers. In Egypt, however, Sirius was not generally connected with these concepts; rather it was associated with the Goddess Isis and the annual flooding of the Nile. None of this, however, quite gets to the point of how Sirius came to be labeled as reddish. The most important astronomical event associated with Sirius was its heliacal rising described in the first chapter. The heliacal rising of Sirius occurs on approximately the 20th of July, depending on the latitude of the observer and historical epoch in question. In the Nile River valley of 3000 to 5000 years ago, the approximate date of this phenomenon would have been the 16th to 17th of July and would have closely coincided with the annual flooding of the Nile, an event of critical importance to Egyptian agriculture. In many parts of the Hellenic world this was also the beginning of the period of intense summer heat. Therefore, just as the sun noticeably reddens on the horizon, it was a similar reddish appearance of Sirius, low on the eastern horizon, just before sunrise that drew the attention of the ancients.

Dr. Whittet also gives perhaps the most compelling astronomical and physiological explanation of the references to a "Red Sirius". He begins by taking Ptolemy and his cohorts at their word—that Sirius appeared red—and inquires under what circumstances would it appear red to an observer? It is well known that when the sun is on the horizon at sunset or sunrise it appears both dimmer and redder. This also applies to bright stars such as Sirius near the horizon. The effect of the scattering of light by the molecules in the earth's atmosphere acts to both shift the color balance of the light from a star and to extinguish the light. Unlike interstellar dust, however, atmospheric reddening is relatively much stronger so that the amount of atmosphere required to make Sirius appear as red as Arcturus or Betelgeuse will only dim the brightness of Sirius by about an entire magnitude or a factor of 2.5. This occurs within 5° of the horizon, at which point Sirius is still one of the brightest stars in the sky and its reddish color can still be perceived by the human eye. Given the ancients were primarily concerned with observing Sirius at heliacal rising, when it would appear much more reddish than when it is higher in the sky, then it is perhaps understandable how the color red might have become associated with Sirius. Later in the year, when the summer heat diminished, the common appearance of Sirius, high in the night sky, was bright blue–white and was not as remarkable.

The author has several times personally observed both the rising and setting of Sirius in attempts to notice reddening. One night during an observing run at Kitt Peak Observatory in Arizona, I carefully watched Sirius rise above the eastern horizon, just after midnight. As the star rose it scintillated violently, taking on a range of colors from an iridescent red to a brilliant blue, sometimes almost disappearing completely, and very much resembling the flashing lights on a police car. Sirius never appeared fully red or reddish. The only time I have ever witnessed Sirius take on a reddish appearance was when it was setting in the west and I was able to follow it with binoculars. Only after it had diminished several magnitudes below naked eye visibility, did it appear reddish. Granted these instances occurred during very clear nights and I was at an elevation of 2000 meters and was able to observe Sirius very near the horizon. Viewing the same events from sea level, through a murky horizon, may well have resulted in a different appearance.

There exists another persistent story that has become a firmly established fixture of the pseudoscientific lore surrounding Sirius, as well as the subject of serious scientific scrutiny. This is the possible existence of a third body orbiting within the Sirius system, Sirius C. There is nothing to preclude the existence of such a faint companion to Sirius. Indeed, there are many examples of stellar systems having three, or more components. For example, just recently Professor Edward Guinan and Ignasi Ribas at Villanova University have discovered an unseen third body in a wide 30-year orbit about the Sirius-like system V471 Tauri. The existence of this particular star, which may be a brown dwarf, was inferred from the systematic changes in the times of the eclipses of the white dwarf by the K2 dwarf primary.

From time to time, there have been reports from visual observers of faint stars seen near Sirius A and B. There was a flurry of such reports during the 1920s, when Sirius A and B were well separated. The first report came from Philip Fox observing at the Dearborn Observatory, in Evanston, Illinois, with the original $18\frac{1}{2}$-inch lens with which the Clarks first observed Sirius B, 58 years earlier. In a long list of Sirius observations covering the period 1915 to 1925, Fox made a brief note next to his observation of February 9, 1920, "B appears possibly double in 231°:0.8″." Essentially, he thought he saw a faint star 0.8 arc seconds southwest of Sirius B. None of Fox's other observations of Sirius reported any third stars. The next reports of a possible third star came in 1926 from R. T. A. Innes of the Union Observatory in Johannesburg, South Africa. In a letter to the British publication *Observatory* he reported: "Images being found to be unusually steady on the evening of 1926 February 4, I examined Sirius with the 26½-inch Grubb refractor. A faint star, about 12th magnitude was seen near the 8th magnitude companion of Sirius, and on the further side. This new companion was seen on that evening by all the available astronomers at the Union Observatory and was measured by Dr. van den Bos." There followed a list of nearly a dozen measurements made by van den Bos, which spanned the period 1926 to 1928. All reports corresponded to a star seen approximately 1.5 to 2 arc seconds to the east of Sirius B. There followed a letter from W. H. van den Bos in which he stated: "May I add a few remarks to the preceding note on a suspected companion to Sirius B? With respect to my measures of this object I feel in much the same position as to those of the companion to Procyon. I cannot feel confident that I am measuring an actual star-image; at the same time I cannot convince myself that I am the victim of an optical illusion." Between 1926 and 1928 the small star appeared to move in a northerly direction with respect to Sirius B. There followed further reports of a third star in 1929 from van den Bos and another observer, R. Finson. After that, no further sightings of the star were forthcoming.

During the 1930s these isolated sightings of a faint third star from North America and South Africa prompted two detailed studies of the orbital motion of Sirius B as a function of time. The question examined was whether there existed any significant periodic departures from the two-body orbit that might indicate gravitational perturbations due to a third body. Both studies pointed to small periodic residuals with a period of somewhere near 6 to 6.3 years. Scientific interest in the possible existence of a third body reemerged in the late 1980s when the issue of the Red Sirius was resurrected. One possible explanation for Sirius having appeared red in the past involved the interaction of Sirius A with a possible third body. The question of the existence of such a body was taken up in 1995 by Daniel Benest of the Nice Observatory in France and his colleague J. L. Duvent. These authors carefully analyzed all the observations of the Sirius A–B orbit up to 1979. They too found evidence of a small perturbation with a period of 6.05 years and with amplitude of 0.055 arc seconds. They were further able to estimate a maximum mass for a hypothetical star corresponding to Sirius C at 0.05 solar masses. This mass would correspond to a body below the hydrogen-burning limit for a proper star (about 0.07 solar masses) and in the mass range of a brown dwarf.

What the Benest and Duvent analysis did not reveal was whether this body orbited Sirius A or Sirius B. To determine which star was most likely to harbor the third body, they conducted a theoretical analysis of the possible stable orbits for a low-mass star around Sirius A or Sirius B. Stable orbits are those which can persist over the age of the Sirius system and which do not result in the ejection of the smaller star from the system. Benest and Duvent found that stable orbits with periods up to six years did exist around Sirius A, but not Sirius B. There are no stable orbits around the less massive Sirius B which exceed three years. Therefore, if Sirius C exists, it must orbit Sirius A. It is also possible to conclude that such a star could in no way be responsible for the flurry of sightings from the 1920s, it would be too faint and too close to Sirius A to have ever been seen by visual observers.

The Benest and Duvent finding of a small perturbation in the Sirius A–B orbit is at odds with the results of two earlier studies. In 1973 Dr. Irving Lindenblad of the U.S. Naval Observatory examined his precise photographic observations of the orbital motion of Sirius B and found no evidence for the presence of a third body. Lindenblad's limit, however, was at about the same level as the amplitude of the perturbation reported by Benest and Duvent. A more definitive result came from the 1978 analysis of 60 years of precise photographic position measurements of Sirius A by Dr. George Gatewood and his wife Carolyn. They also found no evidence for a third body in orbit about Sirius A. Given the precision of their results they should have easily detected a perturbation less than half that reported by Benest and Duvent. At this point in time the Benest and Duvent results remain without confirmation.

It now appears that sightings of "Sirius C" can be attributed to a faint field star that Sirius passed near during the 1920s. We know this thanks to the careful work of Jean-Marc Bonnet-Bidaud and colleagues in France who obtained deep electronic images of the sky near Sirius with a telescope and camera designed to suppress the scattered light from Sirius A. From these images they were able to establish that Sirius did indeed pass near a 12th magnitude field star in the late 1920s and the passage coincided with the timing and location of the Sirius C sightings. We know this star has nothing to do with Sirius because it does not share the proper motion of Sirius A and B, and is now observed to be located more than one minute of arc away.

A second aspect of the Bonnet-Bidaud paper dealt with the presence of any faint stars that could actually be associated with the Sirius system. They were able to rule out the existence of any faint companions down to magnitude 17 in a region extending in radius from 30 arc seconds to 2.5 arc minutes from Sirius A. Essentially they found no stars associated with Sirius down to the substellar brown dwarf limit in this zone. Due to scattered light they could not search closer than 30 arc seconds from Sirius A and thus these results had little relevance to the hypothetical companion of Benest and Duvent.

To get in closer to explore the region near Sirius A, it is necessary to use resources such as the *Hubble Space Telescope*. This was done in 2000 by two groups of observers using two different techniques and instruments. Daniel Schroeder and colleagues used the *Wide Field Planetary Camera 2* on the *Hubble Space Telescope* to search for faint companions around 23 nearby stars, including Sirius. Although they easily saw Sirius B, they found no other companion within 3 arc seconds of Sirius A but could not exclude the possibility that the putative Benest and Duvent star could have been nearer Sirius A at the time of their observations in 1997. Likewise, Marc Kuchner and Michael Brown at the California Institute of Technology searched Sirius, Procyon, and Altair for faint companions using the *Near Infrared Camera and Multi-Object Spectrometer* (*NICMOS*) infrared camera on the *Hubble Space Telescope*. Cool faint stars are more readily detected at infrared wavelengths and these authors were able to rule out the existence of stars down to the mass range of brown dwarfs at 3 arc seconds from Sirius B. They concluded that if a third body does exist then it must be less massive than 40 Jupiter masses. Again there was a small probability that the star was inside of 3 arc seconds at the time of the observation in 1999. It thus appears that there is very little room left for an object of any significant size to orbit Sirius A.

It is possible to go well beyond the optical and infrared searches conducted so far for a possible Sirius C. There is currently an ongoing effort to use the *Hubble Space Telescope Wide Field Camera* to make exquisitely precise measurements of the position of Sirius B. These observations, which extend back to 1997, will be complete in 2007. When complete it will be almost fifty times more sensitive than the Benest and Duvent results and provide unambiguous evidence of the gravitational perturbations of any body down to the mass of Jupiter if it orbits either Sirius A or Sirius B. In Chapter 12 we will discuss these *Hubble Space Telescope* observations that have the potential to either conclusively establish or clearly rule out the presence of such a body.

While there remains a slim scientific justification for a possible Sirius C, it is fair to say that it has yet to be detected. Indeed, existing observations place very stringent limits on how massive such a body can be and where it could be orbiting and still escape detection. Interest in a Sirius C is not just scientific. As we will see much of the popular lore of Sirius, in order to retain any validity, seems to demand the existence of such a body, given that the stellar surfaces of Sirius A and Sirius B themselves are not credible abodes for extraterrestrial beings. For those committed to such beliefs, there must exist some nonstellar body or bodies that orbits one or both stars.

11

Modern Mysteries

A single star beamed down
its crystal cable
and drew a plough through the earth
unearthing bodies clasped together
couples embracing
around the earth
They clung together everywhere
emitting small cries
that did not reach the stars

Lawrence Ferlinghetti (1919–),
The Plough of Time, 1984

The idea that the stars—or, in particular, hypothetical planets orbiting distant stars—might harbor extraterrestrial life is not new. People have speculated for many centuries, both seriously and fancifully, on the existence of such celestial populations. Giordano Bruno was burned at the stake in Rome in the year 1600, in part for adhering to such speculations. Several generations later, in the more intellectually tolerant climate of Holland, Christiaan Huygens also speculated about inhabited worlds in his *Cosmotheoros*. As a general proposition, specific individual stars were seldom emphasized as the source of extraterrestrial civilizations. Sirius, however, being the brightest and most universally recognized star, has received more than its share of attention in this regard. As far back as 1752 the French essayist, dramatist, and writer, Voltaire wrote a short satirical book entitled *Micromégas*, in which he described a giant being from one of several planets that circle the star Sirius. The "Sirian", as Voltaire referred to him, was of colossal size, some 20 miles tall, and had a life span of one million years. Possessing a philosophical nature and a superior knowledge of an abundance of things, this giant being set out to visit our solar system. On the planet Saturn, he encountered a "dwarf", only one-twelfth his height. Together these two supersized beings set off to visit the other planets of the solar system, finally arriving at earth. At first they imagine the earth to be uninhabited, but soon learn of the presence of humans, so small that the enormous beings can only view them through microscopes. The literary device of the towering Sirian allowed Voltaire to play with traditional human ideas of scale, time, and above all, cultural development. *Micromégas* is a tall tale, clearly in the spirit of Swift's *Gulliver's Travels*, and Sirius is little more than a conveniently familiar star from which Voltaire launched his barbs at 18th century society.

Given the prominence of Sirius in the night sky, perhaps it is to be expected that a number of widely accepted and elaborate scenarios have developed around the star and its influence on past and present civilizations. Much of this popular interest can be traced to the importance of Sirius in ancient Egypt. Ancient Egypt and its religion, myths, funeral practices, and monuments have always been a magnet for those seeking spiritual enlightenment through the recovery of "lost knowledge". There is no need to look very far for examples of mystical belief systems involving Sirius: UFO groups, the internet, and the New Age sections of book stores abound with them. Indeed, the star Sirius provides an excellent example of how today's persistent cultural myths no longer spring from the lyres of poets and bards but from the word processors of an electronically connected world. There are three prominent, well-established examples of Sirius-related beliefs that have managed to attract both popular and academic interest: all in one way or another involve the interaction between beings from Sirius and the earth. The first is the intriguing story of a West African tribe and their possession of seemingly specific knowledge related to Sirius. The second is a popular book that postulates that the earth was visited by aquatic beings from Sirius some 5000 years ago. The third is an apocalyptic cult that developed a tragic fascination with Sirius and the possibility of escaping earth and traveling to the star itself.

One of many obstacles facing those who believe in UFOs, or who hold that the earth has been visited by aliens in the past, is the embarrassing lack of any physical evidence. As professor Carl Sagan pointed out, there exist no clear unambiguous examples of unexplainable artifacts that can only be due to an advanced civilization. Although more of a cultural than a physical artifact, the stories of the Dogon regarding Sirius have perhaps come as close to providing "evidence" as anything else.

The Dogon inhabit a stark, dramatic, semiarid plateau in the West African nation of Mali, due south of the fabled city of Timbuktu. There, approximately half a million Dogon and related peoples live in scattered tribal villages near the edge of the central plateau, where they subsist by farming the soil which collects in the narrow ravines that cut into the plateau. The principal crop is millet, which they store in distinctive tall cylindrical granaries with peaked conical thatched roofs (Figure 11.1). The Dogon and their kin speak a series of related but distinct languages and have managed to maintain a rich indigenous culture and religion that has long resisted Islam and Christianity.

Beginning in 1931, and continuing for over the next two decades, the ethnology and belief system of the Dogon were intensively studied by the renowned French

Figure 11.1. Dogon granary (courtesy Lacane).

anthropologist Marcel Griaule, who was joined later by his colleague Germaine Dieterlen. In an article entitled *A Sudanese System of Sirius*, published in 1950 in a French journal of African anthropology, they described for the first time an elaborate set of Dogon beliefs regarding Sirius, and a most curious companion star. (The "Sudanese" designation refers to the former colony of French Sudan, which at the time included Mali.) There followed a book in 1965 entitled *Le Renard pâle*, or The Pale Fox, that further elaborated on the Dogon beliefs, which was completed by Dieterlen after Griaule's death in 1956. The material relating to Sirius constitutes only a small part of a very extensive Dogon creation mythology described by Griaule and Dieterlen. Nevertheless, these Sirius-related beliefs are startling, to say the least, and have become the staple of much speculation, both popular and scientific, regarding their origin and meaning.

The essential elements of the Dogon beliefs about Sirius are outlined in the 1950 paper and derive from information mostly obtained from the extensive oral interviews of four individuals that Griaule conducted between 1946 and 1950. The principal informant was Ongnonlou Dolo, a village patriarch, who was between 60 and 65 years old at the time. The other three individuals held priestly positions and served mainly to confirm the information from Ongnonlou, which was regarded as a secret body of tribal knowledge. The following contents of the story are of particular interest:

- The star Sirius or *sigu tolo*, the star of sigui, is orbited or circled by a smaller companion called *po tolo* or the "Digitaria star". (Sigui is a key Dogon festival held very 60 years and Digitaria is derived from the botanical name of the smallest grass seed, *Digitaria exilis*, familiar to the Dogon.)
- The path of *po tolo* is an elongated oval shape, with Sirius occupying not the center, but a position near one of the (elliptical) foci. There are reproductions of a drawing of the orbit, which the Dogon orient perpendicular to the horizon (Figure 11.2).
- The orbital period of the system is 50 years, while *po tolo* rotates about its axis once a year.
- When *po tolo* is near Sirius, Sirius brightens, and when the separation is greatest, Sirius twinkles and may appear as several stars.
- *Po tolo* is the smallest star, and in fact the smallest thing the Dogon can conceive of. At the same time it is massive and made of a "heavy" metal called *sagala*, which is so heavy (when reckoned in donkey loads) that it cannot be lifted by all the people on earth.
- There is a third member of the system, the star *emme ya tolo*, or Sorghum–Female star, which although larger than Digitaria is only one-quarter as heavy. It travels in a larger trajectory, but in the same direction with the same 50-year period.

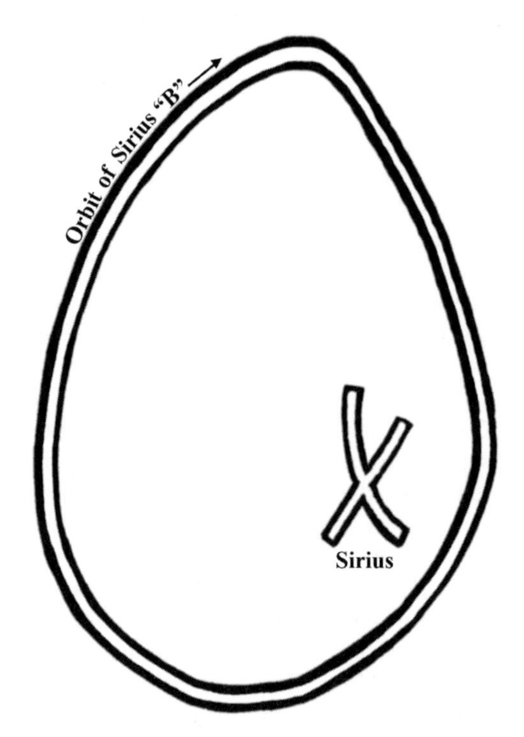

Figure 11.2. A Dogon representation of their concept of the path of "Digitaria" about Sirius (Michael Soroka).

With a few notable exceptions these elements bare a striking resemblance to the information astronomers had acquired about Sirius A and B by the late 1920s. One of these interesting exceptions is that Griaule and Dieterlen report that Sirius is seen by the Dogon as a red star, while Digitaria appears white. This piece of information has obvious parallels to the Red Sirius controversy (see Chapter 10) that was undergoing a revival in the late 19th and early 20th century. The above elements have also become the primary basis for an enormous amount of informed and uninformed speculation on how the Dogon might have acquired such information. It is clear from the content of the stories that Sirius is regarded by the Dogon as the source of life and the embryo of the world and much of the lore is associated with this role. Indeed, the Sirius-related elements are strongly interlaced with the symbolism of the fecundity of women and of nature. The orbit of Digitaria is located at the center of the Dogon world and the star is the axis of the world. There also exists an important, but ambiguous link between the 50-year orbit of Digitaria and a key Dogon ceremony called *Sigui* that is celebrated every *60 years* to mark the renovation of the world. Wooden masks are created for this ceremony, and afterwards are stored in specially built shelters. When Griaule counted these generations of masks, he noted that they marked three or four prior Sigui, indicating that the ceremony had been celebrated for at least several centuries, and perhaps much longer. Griaule and Dieterlen, however, provide no historical information on when or how the Dogon developed their ideas regarding Sirius, leaving such speculation to others. The Dogon, themselves, or at least Ongnonlou, explained that in the distant past the eighth Dogon chief had cheated death by visiting Sirius and learning of its secrets.

Griaule and Dieterlen are the only academic sources for the Dogon Sirius-related material. Virtually all evidence that supports this myth-based Sirius lore is contained in *A Sudanese System of Sirius* and *Le Renard pâle*, and no independent sources or confirmations of these stories have ever emerged. The written source material for the stories was generated by Griaule, who worked alone through an interpreter to conduct extensive interviews with a small circle of Dogon elders. These unique accounts of Dogon beliefs regarding Sirius would have likely gone unnoticed were it not for Marcel Griaule's fame and renown as an anthropologist. The Sirius material appeared shortly after Griaule's 1948 book *Dieu d'eau* (The God of the water) which had a widespread academic and popular impact, particularly in Europe. *Dieu d'eau* revealed a surprisingly sophisticated Dogon view of creation, cosmology, religion, and philosophy. It also helped popularize African art and African design motifs, as well as to introduce elements of African thought into psychology, philosophy, and the history and development of religion. The book also placed the Dogon on the world stage as a prototypical African people, and helped foster further studies of the Dogon and other African groups.

The Sirius stories are so unusual in their seeming astronomical and astrophysical detail and specificity that they have generated an enormous amount of subsequent interest and dispute. Much of the "popular" speculation accepts the stories at face value and seeks to answer the question of how the Dogon might have acquired this knowledge without the aid of telescopes, and the necessary theoretical physics to interpret such observations. There are several major variants of these theories that purport to answer this question. By far the most widespread popular belief is that the Dogon essentially acquired their knowledge from extraterrestrial contacts in the distant past. The most elaborate of these scenarios is the major theme of the best-selling book *The Sirius Mystery*, by Robert Temple, which is discussed in more detail later in this chapter. Another set of explanations, also located in the distant past, assert that the Dogon were in contact with the ancient Egyptian culture and that they are the last remnants of an early sub-Saharan scientific culture that possessed crude optical aids, or simply had superhuman visual acuity. These latter theories are prevalent in Afro-centric literature. However, none of these present day speculations are evident in the work of Griaule and Dieterlen, where the material is presented without comment or relation to the outside world—just as if they were describing Dogon ceremonial dances or agricultural festivals. Nevertheless, any good-faith acceptance of the stories would seem to beg a reasonable explanation, in particular one that does not require ancient mysterious extraterrestrials from a distant star.

Such rational explanations take two basic forms, both of which involve the contamination of Dogon mythology by modern astronomical facts. In one case, the contamination could have arisen during the very process of acquiring the stories themselves; this could be called the "anthropological" explanation, since it comes from some members of the professional anthropological community. A second explanation involves an unknown third party who might have imparted the scientific seeds of the Dogon stories, prior to their having been recorded by Marcel Griaule. This could be called the "astronomical" explanation, since it naturally appeals to most astronomers.

To the international anthropological community, where Griaule commands a considerable reputation, the Sirius stories are but a small part of a much larger debate about his research and methodology. The main academic critic of Griaule's methodology is the Dutch anthropologist Prof. Walter van Beek, from the University of Utrecht. Van Beek worked with the Dogon between 1979 and 1980 and published a paper in which he tried to verify some of Griaule's Dogon material, including the Sirius stories. He was unable to confirm much of what Griaule and Dieterlen had reported about Dogon belief systems. For example, van Beek found the Dogon had no knowledge of Sirius as a double star, and that astronomy seemed to play a small role in their indigenous religion. It is van Beek's view that the Sirius stories must have grown out of a complex interplay between Griaule, his informants, and interpreters. According to van Beek, the stories are not so much a one-sided narrative presented by the Dogon elders as they are the result of in-depth questioning by Griaule, and his interpretation of the elders' responses. In particular, van Beek believes that one informant, Ambara, spoke of Sirius and its companion as belonging to different generations, not specifically as a double star system. Certain stars were pointed out to Griaule in the sky, which the elders considered to be the "father" and "son" of Sirius. Prof. van Beek also believes Griaule was aware, in some fashion, of the astronomical nature of the Sirius system: knowledge that he acquired as a student in Paris. The Sirius stories then were primarily the result of miscommunications between Griaule and his informants, followed by an over-interpretation of the information the elders were providing. Prof. van Beek is careful not to suggest that Griaule fabricated or invented the stories.

Marcel Griaule and his results continue to have many defenders among fellow anthropologists. Chief among these is his daughter, Dr. Geneviève Calame-Griaule, a well-known linguistic anthropologist who has worked among the Dogon, and who was with her father in the field during much of the relevant period when the Dogon stories emerged. Dr. Calame-Griaule has published a detailed rebuttal of van Beek's conclusions. She emphatically states that her father knew nothing of astronomy and had no training or particular interest in the subject. She recalls that her father was unaware of the significance of the Sirius stories until he returned to Paris and inquired about them with André Danjon, director of the Observatoire de Paris. Moreover, she states that it was the Dogon who first spoke to her father about Sirius, and not the other way around.

To most astronomers, the Dogon stories seem to represent some form of cultural transmission. Indeed, it has been argued by Professors Carl Sagan, Kenneth Brecher, and others that the Dogon must have acquired the particular and specific aspects of their beliefs regarding Sirius from contact with knowledgeable European visitors perhaps as early as the 1920s or 1930s, but possibly later. In essence, under this hypothesis, the stories represent a remarkable case of cultural diffusion. This raises a number of interesting questions. First, who might these informed visitors have been: traders, French colonial officials, missionaries, or anthropologists? Second, why would outside visitors wish to discuss recent developments in astrophysics with the Dogon? Assuming that Griaule and Dieterlen were, in fact, the first anthropologists to contact the Dogon, then—as Kenneth Brecher, of Boston University, has suggested—missionaries are plausible guesses for those responsible for the cultural transmission. Missionaries, of all people, would perhaps have the most motivation to discuss such matters with the Dogon. If, as appears to be the case, the Dogon had a set of pre-existing historical religious or cultural beliefs regarding Sirius, then missionaries may well have inquired into those beliefs in the process of trying to convert the Dogon. Upon learning that Sirius held an important place in Dogon cosmogony, a scientifically literate missionary may have tried to enlighten the Dogon about the present state of scientific knowledge regarding the star, in the hope of demystifying it. Indeed, the primarily Baptist-oriented Sudan Evangelical Mission was established among the Dogon in 1931, the year of Griaule's first field trip to the region. Partial translations of the Bible into the Dogon dialect began to appear as early as 1933, clearly demonstrating that contact with religiously-oriented Europeans had been firmly established.

All of the Dogon's claimed unknowable knowledge of Sirius described by Griaule and Dieterlen was in the public domain in Europe by 1926. The subluminous nature of the white dwarf was becoming apparent in the period 1910 to 1915. The high mass and the elliptical orbit were known for fifty years before that, and the reported sightings of the third body (see Chapter 10) date from the early 1920s. The critical demonstration of the high density of Sirius B came in 1925 from the observations of Walter Adams. However, little of this information would have been widely circulated in popular literature, were it not for the fact that it was coupled with the third and final classical test of Einstein's theory of General Relativity, the 1925 observation of the gravitational redshift (Chapter 9). By that time, Einstein was already a well-known public figure, and his theory of General Relativity, and the observations that confirmed it, were widely reported in the press and in popular science publications of the period. For example, the *New York Times* ran three stories on Sirius B in 1925 (see Figure 11.3), and the *Times* (of London), published one. Brecher, in his discussion of this question, also notes contemporary reports in *Scientific American* and the paper *Le Monde* in Paris.

Arguably the most ubiquitous source of such knowledge was Eddington's book, *Stars and Atoms*. As mentioned in Chapter 7, this book was based on a series of popular lectures delivered in 1926 and published in 1927. It was intended for, and widely read by, the general public; French and German editions appeared soon after the English version. The book included an essay on Sirius B entitled *The Story of the Companion of Sirius*, subtitled *The Nonsensical Message*, which contained explicit information on the small size and incredible density of Sirius B. This was also accompanied by an appendix which gave even more information, including an estimate of the orbital period of 49 years, a discussion of other known white dwarfs, and Fowler's then recent work (Chapter 8) explaining why such stars are so dense. There was also a simply stated but comprehensive article by Paul Baize on Sirius A, B, and C which was published in 1931 in the *Bulletin of the French Astronomical Society*, which also contains this same information. Moreover, one need not reach all the way back to the 1930s for the potential origin of the Dogon lore. In spite of their long history of working with the Dogon, the stories on which Griaule and Dieterlen's reports were based were collected after the Second World War, in the late 1940s. Thus, given all

NEW TEST SUPPORTS EINSTEIN'S THEORY

Result Is Achieved in Confirming Eddington Star Density Theory at Mount Wilson.

SPECTRUM SHIFT SIMILAR

Prediction for Satellite of Sirius Is Proved True—Analagous to Einstein's for Sun.

Figure 11.3. Adams' gravitational redshift measurement of Sirius B (*New York Times*, July 22, 1925, page 21).

the attention Sirius was receiving at the time, it is not entirely out of the question that a missionary or someone else, with a basic knowledge of contemporary astronomy and physics, could have discussed these matters with the Dogon. This idea is supported by the clear echoes of several Biblical and Christian themes that are part of the Dogon mythology described in the Griaule and Dieterlen book *Le Renard pâle*. These include a Dogon tale about a form of crucifixion of the Nommo on a tree and the resurrection and return of the Nommo. The Nommo are aquatic beings involved in Dogon creation mythology. Dr. Calame-Griaule, however, discounts these similarities as coincidence, since many cultures have stories of sacrifice, rebirth, and renewal.

How likely is the missionary scenario? The answer is that we simply don't know. John McKinney is the son of the missionary family that started the first mission among the Dogon in 1931, and he still resides in Mali. He is fluent in the Dogon language and has spent most of his life in the area of Mali where Griaule and Dieterlen worked, and was personally familiar with many of the individuals mentioned in *Le Renard pâle*. Although McKinney is familiar with the outside interest in the Dogon stories regarding Sirius, he has no knowledge of any special lore concerning Sirius that is central to Dogon culture or beliefs. Specifically, however, he knows of no one associated with the Protestant mission in which he grew up who could have passed on any scientific knowledge of Sirius to the Dogon; nor is he aware of any other outsiders who were likely candidates to have done so. This is backed up by Dr. Calame-Griaule, who emphatically rejects the idea that missionaries or colonial officials could have possibly been the source of these stories. In summary, in spite of their unique insights, McKinney and Calame-Griaule remain as puzzled as anyone else as to the origin of the Dogon stories. However, even though there presently exists no specific external source of this material, an unknown third party certainly remains a more plausible explanation than the alternatives involving extraterrestrial contacts.

For those who doubt that strange unlikely coincidences do in fact result from contacts between isolated traditional societies and the modern world, I can relate a personal story regarding another isolated tribe in Africa. In 1976 I was working in Kenya, and traveled with a group of expatriate friends into a very remote part of the country, well away from the popular tourist spots. We traveled for hours along isolated dirt tracks into the remote Kerio valley in western Kenya. That evening, we made camp in a very overgrown area off the road when the inhabitants of a small nearby village approached to investigate the strangers. We tried to ask permission to camp but had trouble communicating since the villagers spoke little Swahili, a sure indication of the isolation of these people. I went to the village to speak with the Headman but was still having trouble communicating. Finally he took me to see a "government man", who turned out to be a low-level district officer who was looking into the possibility of constructing an airstrip in the valley. The officer spoke English, Swahili, and the local language and was able to translate between the village Headman and myself. After receiving permission to camp, the Headman learned I was an American. His face lit up and he asked me if I knew Henry Kissinger, the well-traveled Secretary of State for then President Ford. I was a bit startled by the question but thought that he must have picked up the name from a local language broadcast on the

"Voice of Kenya" radio. I replied that I knew of Dr. Kissinger, but of course had never met him. At that point the man nodded exclaiming, "Yes, Henry Kissinger! He is a very good man!" He then vigorously shook my hand saying "I met him and shook his hand like this." This made no sense at all, so I asked the district officer what the Headman was talking about. After some convoluted translations, I learned that a few months earlier the Headman and a troop of village dancers had traveled to Kenyan President Jomo Kenyatta's presidential compound at Lake Nakuru. There they performed for the president and his guest, Henry Kissinger, and members of the village were introduced to Kissinger. At that point, I recalled Kissinger had recently visited Kenya. The Headman was correct. He had shaken hands with Kissinger. Henry Kissinger may not remember meeting the gentleman from the remote Kenyan village, but the Headman certainly remembers Dr. Kissinger.

The point of this digression, apart from problems of communication, is that it is not always sound to make assumptions about what seemingly isolated tribal populations in Africa and elsewhere are and are not aware of regarding the outside world, and that when the unexpected happens, we shouldn't be too surprised. Carl Sagan, in his book *Broca's Brain*, and Guy Consolmagno, in his book *Brother Astronomer*, recount other instances of traditional peoples who have assimilated unlikely bits of contemporary, or even ancient, western culture into their rituals and belief systems, only to later recount them to bewildered outsiders.

There also exists an interesting historical footnote to the story of the Dogon and their beliefs about Sirius that reach back to colonial America. The central character is Benjamin Banneker, who was, among many other things, a surveyor, a clock maker, a writer of almanacs, and an early scientist. Banneker's grandfather was a first-generation West African slave named Bannaky, and his grandmother was an indentured servant from England named Molly. They started a small farm and raised seven children in rural Maryland. One daughter, Mary Bannaky, purchased Benjamin's father Robert as a slave to work the farm. She later married him and together they had a son Benjamin, in 1731. When Robert died, Benjamin was raised by his mother and learned to read and write from his grandmother, Molly. Later, he attended a small rural school for a few years where he discovered an interest in mathematics. By all accounts, he also had an unusual interest in nature and a fascination with all things mechanical. As a young man, after examining the workings of a pocket watch, he built a hand-carved wooden clock that worked for fifty years. He also came into contact with the Ellicotts, a Quaker family that ran a mill in rural Maryland. It was through Andrew Ellicott that Banneker became interested in astronomy and surveying. Banneker was one of the three men who surveyed the original site of the future U.S. capital, Washington, D.C. He was largely self-taught in astronomy, using contemporary textbooks borrowed from the Ellicott family. Between 1793 and 1797, Banneker published a widely read popular almanac based on his own calculations. He also owned a small brass telescope and viewed the stars through a hole he cut in the roof of his rural cabin.

It has been alleged that Banneker had a strong interest in Sirius and that he believed that it possessed a smaller companion. Indeed, Charles Cerami, the author of a 2002 Banneker biography, has linked this belief to Banneker's paternal grandfather, who it is said may have originated from the Dogon tribe. Cerami also mentions Banneker's belief in the possibility that habitable planets circled many stars. This latter belief is in fact to be found in Banneker's 1792 Almanac, but the source for Banneker's reported interest in Sirius, if any, is simply not known. There is, for example, no mention of Banneker's interest in Sirius in the authoritative 1972 biography of Banneker by Silvio Bedinni, nor among what remains of Banneker's writings. Further, Bedinni traces Banneker's paternal grandfather, who was also called Bannaka or Banna Ka, to the Wolof people of West Africa, rather than the Dogon. Unfortunately, we possess little if any first-hand knowledge of what Banneker actually thought about the stars, or Sirius in particular, since much of his writing was destroyed in a fire soon after his death in 1806. The Banneker connection with Sirius appears very much a product of 21st century mythology, perhaps traceable to the assertions in Cerami's biography.

Without a doubt, the best known and most widely circulated of all Sirius-related beliefs is contained in a 1976 book entitled *The Sirius Mystery*, by Robert Temple. The central thesis of the book is that about 5000 years ago the earth was visited by advanced beings from the Sirius system and that this extraterrestrial encounter was the genesis of the ancient Egyptian and Sumerian civilizations. Much of Temple's case for such a visit rests on the Dogon stories. *The Sirius Mystery* was enormously successful, going through three editions in the 1970s and 1980s. A much-expanded edition, which promised "new scientific evidence", was brought out in 1998. The American-born author, Robert Temple, is a writer residing in London and lists on the book jacket a number of scientific and academic societies to which he belongs, including The Royal Astronomical Society, The Royal Historical Society, and the Institute of Classical Studies, among others.

The central mystery, to which the title of the book refers, is the Sirius lore of the Dogon and how they could have acquired such knowledge. In unraveling this mystery, Temple follows a path that leads him back to the ancient civilizations of Egypt and Sumer, and the seeming sudden rise of these cultures 5000 years ago. This and other evidence also lead the author to his conclusion that the earth was visited in prehistory by advanced aquatic beings from Sirius. These extraterrestrials imparted superior knowledge to the ancient world and left vestiges of knowledge about the Sirius star system that was retained by the Dogon. The "new scientific evidence" concerns a possible detection of a third member of the Sirius system that appeared in the scientific literature in 1995. Recall that Dogon lore does identify a third "star" in the Sirius system. Temple regards the 1995 French report of a possible Sirius C (see Chapter 10) as confirmation of Dogon lore and a vindication of his "prediction" contained in the 1976 first edition of *The Sirius Mystery*, that such a star might be found.

Temple begins the book with a summary of the scientific knowledge of Sirius, but attaches undue significance to the specific numerical values of the mass and radius of Sirius B with respect to the sun. In a flourish of numerical mysticism, he freely uses these values to construct a host of coincidences with the dimensions of the Pyramids of Giza and certain "sacred ratios" from ancient studies of musical harmony. The reason such numerical agreements exist, he argues, may lie in the idea that Sirius and the sun are close enough in space to inhabit the same "cosmic cell" of interstellar space. For reasons that will become evident, Temple has even come up with a name for this volume of space: the "Anubis Cell". Within this cell he suggests it is possible for stars and other things to mutually share collective properties.

The specific values that Temple quotes for the mass and radius of Sirius B, 1.053 solar masses and 0.0078 solar radii, are taken from Kenneth Lang's 1992 edition of *Astrophysical Data*. These particular values in turn come from George and Carolyn Gatewood's (1978) study of the orbit of Sirius, which rely heavily on a 1960 analysis of the orbit of Sirius B. In working out his finely tuned numerical coincidences, Temple neglects to include the fact that the scientific estimates of quantities such as the mass and radius of Sirius B almost always come with numerical qualifiers, or confidence limits. In the case of the Gatewoods' estimates, these are 1.053 ± 0.028 and 0.0078 ± 0.0002. The purpose of these confidence intervals is to define a measure of the probability that the estimates lie within the specified interval. The standard convention, which applies to the estimates of the mass and radius of Sirius B, is that there is a 68% probability they lay within, and a 32% probability they lay outside, these limits. For example, just after Temple published his most recent edition of *The Sirius Mystery*, a number of colleagues and I published a revised determination of the mass and radius of Sirius B based on new space-based observations. The new values differ from those used by Temple; we found 1.034 ± 0.026 and 0.0084 ± 0.00025, respectively for the mass and radius. As will become evident in the next chapter, even newer observations from the *Hubble Space Telescope* give a somewhat smaller, but much more precise, determination for the mass for Sirius B. These changes demolish, or at least render much less significant, most of the numerical coincidences cited by Temple with respect to Sirius B.

Likewise, Temple attaches a great deal of significance to the existence of a putative third star in the Sirius system, Sirius C. This is the essence of the "new scientific evidence" promised in the 1998 edition of *The Sirius Mystery*. Temple correctly reviews the scientific case for the existence of such a third star, which was initially prompted by reports from several visual observers in the 1920s, and he even cites the negative result of Irving Lindenblad, who failed in 1973 to find dynamical evidence in his very accurate observations of Sirius. However, Temple seizes on the 1995 paper by the French astronomers, Benest and Duvent, in which these authors reported some dynamical evidence for the existence a third body. The data they examined were best fit by a hypothetical star with a mass less than 5% that of the sun and in a 6-year orbit about Sirius A. As the authors point out, actually detecting such a star would be difficult since it would be quite faint. The entire issue of Sirius C was extensively covered in the previous chapter, but it is now becoming clear from observations made with the *Hubble Space Telescope* that there is likely no star with the properties specified by Benest and Duvent. In particular, bodies such as possible brown dwarfs and planets the size of Jupiter can almost certainly be ruled out. It is always possible to hide a smaller, less massive body in the Sirius system, but the existence of such a body would become almost impossible to detect. In summary, as far as is known the Sirius system consists of just two stars, neither of which is a conceivable abode for life.

The bulk of *The Sirius Mystery*, however, is far more difficult to encapsulate or deal with. In summary, the key that unlocks the *Mystery* turns out to be the Dogon. After dismissing the possibility that the Dogon could possibly have been influenced by 20th century outside contact, Temple seeks a more certain answer in the distant past. In his view, the Dogon represent the only living remnants of a lost 5000-year-old tradition of knowledge that was the driving force in the rise of the ancient civilizations of Egypt and Sumer. The Dogon, he suspects, are the descendants of the Garamantians. The Garamantians and their North African homeland are memorably described by the 5th century BC Greek historian Herodotus, as a land where the native oxen graze backwards. Temple hypothesizes that the Garamantians had contact with the ancient Egyptians and learned their priestly secrets. Much later, under pressure from the Romans, the Garamantians migrated to the south, across the Sahara, and west, into the region of the upper Niger River. There they settled and amalgamated with the local population to produce the Dogon. According to Temple, the Garamantians carried with them the essential mythical elements of the pivotal encounter with extraterrestrials, which survive in the form of the Sirius lore of the Dogon. Thus, this secret knowledge made an exodus from ancient Egypt, crossed the Sahara with the Garamantians, and settled among the Dogon. This hypothetical chain of events effectively transforms Temple's *Mystery* into a new quest for the roots of the extraterrestrial contact among the myths of ancient Egypt and Mesopotamia and the mythology of classical Greece.

To provide a flavor of just one of the many strands of evidence that Temple weaves together, consider Cerberus, the dog of Greek mythology that guarded Hades. Originally, Cerberus was the 50-headed (but more frequently 3-headed) dog that Hercules subdued and carried out of Hades. Cerberus also had a fifty-headed twin, Orthrus. Temple's interpretation of Cerberus/Orthrus tracks the following train of evidence: Orthrus, a dog = Anubis the Egyptian jackal-headed god, who guarded the realm of dead, which in turn was ruled by Osiris, who was the husband of Isis, who was a manifestation of Sirius. Moreover, Cerberus/Orthrus (in some legends) had fifty heads and this is therefore sound evidence that the ancients knew that Sirius B orbited Sirius A in fifty years. Apparently, generations of scholars missed all these clues because they were unaware that Sirius had an unseen companion with a fifty-year period. And how did the ancients know all this and more? It was deeply embedded in "Black Rites" and "Dark Rites" that were the closely-held knowledge of the Egyptian priestly class and which related to the dark star Sirius B. This knowledge, Temple claims, is today only discernible in elliptical references gleaned from ancient texts. The ultimate sources of this knowledge, according to Temple, were of course the "mysterious visitors" from Sirius.

To make the final leap from ancient mythology to advanced visitors from Sirius, Temple relies on both Babylonian and Dogon legends. The Dogon creation myth includes aquatic chimeras called Nommo. Likewise, the Babylonian mythological pantheon contains a human-headed god, Oannes, with a fish tail. Temple provides figures of numerous Mesopotamian carvings showing fish-tailed gods. Indeed, in one Assyrian carving (Figure 11.4), a fish-tailed god stands before a stone altar with a winged eye hovering over it. The winged eye, which Temple calls an "eye-star", is he

Figure 11.4. Temple's figure of the Mesopotamian fishtailed god Oannes the "eye star". The winged god not discussed by Temple is on the extreme left (Michael Soroka).

asserts the "mouth of Nommo." Unmentioned by Temple is a second figure that stands behind the fish-tailed god. This human figure lacks a fish tail but does possess substantial wings. Perhaps the winged being is from another star.

It is through the amphibious nature of the Nommo and Oannes that Temple deduces that the long-ago visitors from Sirius must have been fundamentally aquatic creatures. He imagines that they came from a water-covered planet orbiting the elusive Sirius C and after their work was done here on earth, they departed. Perhaps they returned to the Sirius system, or conceivably they merely retreated to an outpost in the distant parts of our solar system, only to await their opportunity to return—Temple isn't sure. He does, however, make the helpful suggestion that we should be on the lookout for odd moons circling the outer planets. In particular, he mentions Phoebe, a distant moon in a retrograde orbit about Saturn. Unlike some other major Saturnian moons, Phoebe was not well observed during the *Voyager* flybys in 1980. Temple expressed the hope that NASA's *Cassini* mission would have an opportunity to investigate Phoebe when it arrived at Saturn in 2004. On June 11, 2004 *Cassini* flew within 2000 km of Phoebe, not to indulge Temple's baseless speculation, but to take advantage of a unique close flyby opportunity during its approach to Saturn. *Cassini* obtained stunning pictures of an icy pock-marked surface. None of the images revealed any evidence of extraterrestrial constructs or mysterious aquatic beings.

Perhaps one of the most bizarre and tragic associations of Sirius and the occult played itself out across two continents over a three-year period in the mid-1990s. Around midnight on the evening of Wednesday, October 4, 1994, a fire broke out in a farmhouse near the village of Cheiry, in western Switzerland. When firemen arrived, they discovered the body of a man lying on a bed. He appeared to have been murdered and his head covered with a plastic bag. The police were summoned and soon discovered incendiary devices and, ominously, the personal belongings of several people in a nearby garage. Although the farm buildings seemed deserted, the garage was found to contain a false wall that led to a hidden underground room that had been converted into a bizarre sanctuary with a large mirror and crimson cloth covering the walls. In the room, police found 22 additional bodies arranged in a circle with their feet pointed inward, some wearing capes of white, gold, and black fabric. The heads of most of the victims were wrapped in black plastic garbage bags, many had been shot in the head, others may have suffocated. The victims, who included a 10-year-old boy, appeared to have been injected with a powerful drug.

That same night, three hours later, more fires broke out, this time in another small Swiss village 168 km to the east. The location was a cluster of three houses in the tiny village of Granges-sur-Salvan. When police arrived, they discovered two of the houses on fire, and inside the bodies of 25 people who had been shot or suffocated, with their heads in black garbage bags. In all, 48 people were dead in what appeared to be two separate inexplicable mass murder–suicides. Perhaps just as disturbing as the events of that evening was the identity of some of dead: a multimillionaire Swiss sales manager for Piaget watches; a wealthy 72-year-old Swiss landowner; a middle-aged accountant from the Quebec Ministry of Finance; a journalist from the *Journal de Quebec*; and the mayor of the town of Richelieu, Quebec, and his wife. The one common bond among the dead at both locations was that all were members of a secretive group called The Order of the Solar Temple. Later it would also be learned that among the dead at Salvan were two members of a clique within the Solar Temple, Joel Egger and Dominique Bellaton, who had returned just days before from Montreal, Canada, via Zurich, after conducting the bizarre assassination of three "traitors."

The events that set in motion the deaths in Switzerland began four days earlier on Sunday, October 1, at a neat green chalet in the small resort town of Morin Heights, Quebec, north of Montreal. Joel Egger, a 34-year-old Swiss citizen, had flown into Canada the previous day from Zurich. The chalet's absentee owner, a reclusive Zurich-based spiritualist, had dispatched Egger to Montreal on a mission. Egger immediately drove to Morin Heights where he was greeted by the Genouds, the husband and wife caretakers at the chalet, and by Dominique Bellaton. Bellaton, an attractive 36-year-old divorcee, was the former mistress of the chalet's owner. The four were soon joined by three others, a Swiss handyman named Antonio Dutoit, his British wife Nicky and their three-month-old son, Emmanuel. The Dutoits had come to the chalet to visit Dominique Bellaton, Nicky's closest friend, who frequently babysat for the Dutoit's son. Antonio was called to the cellar by Egger to help move some furniture. When Antonio reached the foot of the stairs, Egger grabbed a baseball bat and bludgeoned him and then cut his throat with a kitchen knife. Quickly moving upstairs, Egger, joined by Jerry Genoud, then repeatedly stabbed Nicky. Next, the infant, Emmanuel, was ritualistically stabbed twenty times in the chest and left in a plastic bag with a wooden stake driven through his chest. Egger and Bellaton, their work done, caught the next flight for Zurich, while the Genouds cleaned up the chalet. On Wednesday, twelve hours before the events in Switzerland, the Genouds took powerful sedatives, set fire to the chalet and perished in the blaze.

By Saturday, October 7, police in Switzerland and Canada were focusing on trying to locate two men, Joseph Di Mambro, a 70-year-old French national and owner of the Morin Heights chalet, and a 47-year-old Belgian doctor, Luc Jouret.

Di Mambro and Jouret were the spiritual leaders of the Order of the Solar Temple. Di Mambro, a former clock maker and jeweler from southern France, had for decades been associated with a variety of fringe religious groups. By the early 1970s, Di Mambro had founded several groups that were successful in attracting wealthy individuals eager to pay large sums for initiation into the secrets of occult knowledge. What Di Mambro was preaching was an eclectic mixture of Masonic ritual and Rosicrucian secret knowledge, all updated with popular forms of spiritual enlightenment and modern environmentalism. In the mid-1980s Di Mambro joined forces with Luc Jouret, a man with a magnetic personality and charismatic speaking style. Together, they founded the Order of the Solar Temple in 1984. Luc Jouret quickly became the public face of the Solar Temple. He had received a medical degree but soon drifted into alternative medicine and, during the 1980s, built thriving homeopathy practices in France, Switzerland, and Canada. He also began a second career as a motivational speaker lecturing on naturopathy, natural foods, and the environment. Jouret used his public lectures on these themes to recruit new members in bookstores and at student-sponsored university meetings. Interested attendees would be invited to join an inner circle where more overt Solar Temple doctrines were discussed. Apparently, new Solar Temple members would then be encouraged to join ever more exclusive levels of the group. Di Mambro and Jouret slowly built a devoted following of several hundred in France, Switzerland, and Canada. A surprising number of successful professional and business people were attracted to the organization. For example, 15 managers and executives of the giant public utility Hydro-Quebec were recruited after Jouret was hired by the company to give a series of motivational talks. Potential members were encouraged to seek further enlightenment at special seminars. By the late 1980s there were perhaps 400 active members in Europe, Canada, and the French Caribbean. Significant attention was paid to the personal wealth of prospective members. Committed Solar Temple members were encouraged to unburden their lives by contributing property and cash, much of which went toward the purchase of exclusive properties in Europe and Canada and to support the lavish life styles of Di Mambro and Jouret.

The beliefs of the Solar Temple contained several familiar strains. Di Mambro spent many years as a member of the Rosicrucians, a group with secret rituals and the self-appointed custodians of esoteric knowledge passed down from the ancient Egyptians. He also borrowed heavily from the beliefs and rituals of various European Templar societies. The Knights Templars were one of several monastic–military orders established to protect Christian pilgrims in the Holy Land during the Crusades that somehow emerged as one of the more enduring institutions of that period. They grew powerful during the 13th and 14th centuries, rivaling the emerging European monarchies, before King Philip IV of France decimated them in 1307. Since that time, secretive, Templar-like movements and groups have frequently reemerged throughout European history. These groups are often characterized by a strict hierarchy, sworn allegiances, elaborate rituals, and a jealously protected body of secret knowledge. The Solar Temple borrowed heavily from the medieval ceremonial repertoire of the Temple movements. Initiated members participated in dimly lit secret rites presided over by richly robed senior leaders. To this was added a belief that members were reincarnations of famous historical figures. For example, Jouret variously claimed to be the third Christ as well as the reincarnated Osiris and Moses. Di Mambro's mistress, Dominique Bellaton, was proclaimed as the reincarnation of the Egyptian queen Hatshepsut. Members were encouraged to believe that they were special beings, destined to play a crucial role on earth.

Perhaps the most distinguishing characteristic of the Solar Temple was their fascination with the star Sirius. It was a cardinal belief of the group that they would eventually leave their earthly bodies and travel to Sirius. Some 26,000 years ago something they referred to as the "Blue Star" had deposited on the earth beings called "Sons of the One". The Blue Star was a rather murky concept, somehow related to energy from Sirius. (The 26,000-year time span is likely the precession period of the earth's pole, although the Temple members would have thought of it as the cycle of zodiac ages.) Temple members believed that in the early 1980s spiritual forces had requested the "Blue Star" to intervene and forestall the inevitable "changes", which were imperiling the earth. Seven additional years had been granted and that grace period was now coming to an end. The Blue Star was supposed to reappear in the sky and an event involving a gathering of the "Sons of the One" was imminent. The Blue Star would "magnetize" the elect believers, gather and transport them back to Sirius, where they would inhabit glorious "solar bodies". These beliefs may have originated with Di Mambro, who is said to have held a fascination with what he believed was a pre-Christian gnosticism that originated in ancient Syria and Canaan. He thought of Sirius as a refuge where one's divine soul could travel to escape the decadence and decay of the terrestrial world.

Among the secret Solar Temple ceremonies, conducted in a darkened inner sanctum, were elaborate sound and light productions, in which members would listen to sounds from Sirius and watch shimmering images of swords and chandeliers appear in mid-air. This would be followed by the appearance of the "Masters of Zurich". The Masters of Zurich that Di Mambro claimed to represent were a group of supernatural beings that resided in a subterranean abode beneath Zurich. Di Mambro also spoke of a thirteen-year cycle that was begun by the Grand Lodge of Sirius (or of "Agartha") in 1981 and was to be complete in 1994. There was additional mention of *la Grande Loge Blanche de Sirius* (The Great White Chapter of Sirius) composed of the highest ranking spiritual beings who would one day summon the "authentic bearers of ancestral wisdom" from the earth. The mass suicides were envisioned as the point of departure for Solar Temple members, the bearers of ancestral wisdom, who were "transiting" on "death voyages" to be reborn on Sirius.

The Solar Temple began to experience a number of crises in the early 1990s. Disillusionment and dissension developed among some key members. Di Mambro's adult son Elie, who was being groomed to succeed Di Mambro, had a falling out with his father—reportedly after Elie accidentally discovered a closet holding the props used in the mystical rites. In addition, Di Mambro was having difficulty with his daughter Emmanuelle, who had been designated the "Cosmic Child" and who was to be among those who were to found a new world. At age 12 the girl was becoming reluctant to assume this role. Emmanuelle was the daughter of Di Mambro and Dominique Bellaton, who became his mistress when she was in her 20s. The members were told Emmanuelle was the result of an immaculate conception. Around the same time there were also financial and criminal probes of the group, as well as unwelcome attention from the press. All of these developments were sowing doubt and causing memberships to dwindle. Perhaps the most significant defection was Antonio Dutoit, who had been involved in the technical setup of the special effects that Di Mambro and Jouret used to astonish members during the elaborate ceremonies. Dutoit had told other members of these things, and he and his wife had left the group. The Dutoits had also earned the personal animosity of Di Mambro by defying his orders that the couple remain childless. Nicky became pregnant and they named their boy Emmanuel, the masculine form of the name of the "Cosmic Child". Di Mambro regarded the name as blasphemous, labeling the three-month-old boy as "the Anti-Christ." It apparently was Dutoit's defection and the presence of the child that prompted Di Mambro to dispatch Joel Egger to join Dominique Bellaton at Morin Heights and kill the family. This rising level of paranoia caused Di Mambro and Luc Jouret to initiate the departure without waiting for the appearance of the Blue Star, and the call from the Grand White Chapter of Sirius.

The deaths in Switzerland a few days later occurred in a climate of increasing apprehension within the group. The suicides were apparently planned as a final act and departure rite for the cult members on their "transit" to Sirius. For the most committed, this resulted in a voluntary group suicide ceremony. Others, who were not as committed, apparently required help, and were either drugged or physically subdued. The fires were part of a ritual cleansing that preceded the transit to Sirius. Police theorize that at Cheiry, two members, using a pistol, made certain all were dead, set the fire and then drove to Granges-sur-Salvan, since the same gun used at Cheiry was later found at Granges-sur-Salvan. Forensic evidence indicated that some of the victims might have resisted there also. The doors to the buildings at Salvan were found nailed shut from the outside. Among the dead at Salvan were Joseph Di Mambro, his son Elie, his wife Jocelyne, and Di Mambro's 12-year-old daughter, Emannuelle. The charred bodies of Luc Jouret, Joel Eggers, and Dominique Bellaton were also found at Salvan.

The tragedies of October 1994 were not the end. A year later, on December 23, 1995 at a remote site near Grenoble, France, police discovered the charred bodies of 16 more Solar Temple members. All had died of asphyxiation, poison, or gunshot wounds. Significantly, the bodies were arranged in the figure of a star. The deaths had occurred sometime earlier and some of the victims had left notes stating their belief and desire to leave the earth and visit another world. Notable among the Grenoble victims were the wife and son of the 1960 French Olympic skiing champion and maker of high fashion eye wear, Jean Vuarnet. This was followed on March 23, 1997 by the fiery deaths of five more Solar Temple members in St. Casimir, a village west of Quebec City, Quebec. The victims were from two families, but—unlike the other occasions—there were survivors. Three teenage children from one of the families who had resisted the suicide rites were allowed to take sleeping pills and remain in a separate building that was not destroyed. This final act of the Solar Temple was ultimately overshadowed by a chillingly similar occurrence in Rancho Santa Fe, California, just three days later on March 26, 1997. Unlike the Solar Temple, the

39 members of the Heaven's Gate cult that committed suicide that day were departing, not for Sirius, but the then recently discovered, comet Hale–Bopp.

In the end, Di Mambro, Jouret, and their accomplices were all dead and there seemed no one left to be held accountable for what had happened. However, attention soon focused on the unlikely person of the then 58-year-old Michel Tabachnik, the former conductor of the Canadian Opera company, and a well-known international musician and conductor. In 2001 he faced a criminal prosecution in Grenoble, France, where many of the victim's families lived. Tabachnik was charged with involvement with a criminal organization, the Solar Temple. Tabachnik had been an early associate of Di Mambro but claimed innocence, saying that he had left the Solar Temple well before the events of 1994 and had no knowledge of the circumstances leading up to the deaths. However, Tabachnik's wife was among the victims at Cheiry. Although Tabachnik was acquitted by a magistrate, the trial brought to light many aspects of the Solar Temple and the conduct of its leaders. The total number of deaths associated with the Order of the Solar Temple had finally reached 74.

Part Five

A Contemporary and Future Sirius

Over the past three and a half centuries Sirius has taught astronomers and physicists many valuable lessons about the stars and the physical universe. All these lessons resulted from difficult observations made with some of the best ground-based telescopes available at the time. In the last fifteen years, however, it has been possible to extend our studies of Sirius and its companion into outer space. With the help of these observations—at new wavelengths and with greater clarity—it is now possible, with some confidence, to describe how the Sirius that we see today, came to be, and what will become of it in the distant future: and how the same physical processes will shape the fate of our sun and the earth.

12

A View from Space

"There are two kinds of light—the glow that illuminates, and the glare that obscures"
—James Thurber (1884–1961)

The author was first drawn to Sirius over twenty years ago while working on the *Voyager* mission that explored the outer solar system. At that time one of my jobs was to find interesting astronomical objects to observe with the spacecraft's ultraviolet spectrometers, as the two *Voyager* spacecraft traveled between the planets. The ultraviolet spectrometers (one on each spacecraft) were rather small instruments, about the size of a shoe box, and were designed to observe the upper atmospheres of the giant outer planets in extreme and far ultraviolet light. In spite of their small size, these spectrometers were quite unique at the time in also being able to observe stars at these very short wavelengths. I would select hot stars at which to point the spectrometers, in order to obtain spectra in the 500 Å to 1700 Å region. Since many of these stars had never been observed before in the far ultraviolet, the *Voyager* spectra were of some considerable scientific interest. After a while I gradually became familiar with the appearance of the spectra of various types of stars at these wavelengths.

In 1983, when *Voyager 2* was traveling between Saturn and Uranus, I decided to target Sirius. I had earlier observed the star Vega, which is nearly identical to Sirius in spectral type, and so I had every expectation that Sirius would appear as just a brighter version of Vega. When I first saw the spectrum of Sirius, I was initially surprised to see a prominent bump in the flux between 1000 Å and 1150 Å, that was not present in the spectrum of Vega. A brief moment of reflection produced the realization that the small *Voyager* instruments were picking up the ultraviolet spectral signature of Sirius B, and that at these short wavelengths the tiny white dwarf was actually outshining the vastly larger Sirius A (Figure 12.1). This shouldn't have really been a surprise, since Malcomb Savedoff, Hugh Van Horn, and François Wesemael from the University

Figure 12.1. Two views of the Sirius system showing the reversal of the relative brightnesses of Sirius A and B at different wavelengths. On the left is an optical image of Sirius A and B taken near greatest separation in 1976 (see cover image). The faint "rays" emanating from Sirius A are instrumental diffraction spikes. On the right is an X-ray image obtained by the *Chandra X-ray Observatory* of Sirius A and B taken in 2001, again the rays are instrumental diffraction spikes. The *Chandra* image has been rotated and scaled to match the separation and orientation of the two stars in the optical image.

of Rochester and colleges had used the *Copernicus* satellite to actually make a similar observation of Sirius several years earlier. Nevertheless, I was intrigued enough to spend some time analyzing the Sirius spectrum and eventually came up with a new temperature for Sirius B of less than 27,000 K. This represented a substantial refinement of the *Copernicus* temperature estimate of $27{,}000 \pm 6000$ K and a large downward revision from the earlier estimate of 32,500 K that had resulted from the Greenstein, Oke, and Shipman observations of Sirius B obtained with the Palomar 200-inch telescope and published in 1971 (see Chapter 9). My colleagues and I reported those observations in the astronomical literature in 1984, but my interest in Sirius and its strange companion persisted.

There are two absolutely fundamental things that we desire to know about Sirius B. What is its precise mass and what is its precise radius? By the late 1980s astronomers already knew the approximate values of these quantities to within an accuracy of 4 to 5%, thus it is the word "precise" that is important. As a general rule truly accurate estimates of stellar masses and radii are difficult to determine, chiefly due to the problems of measuring stellar distances. At the point where the uncertainties in the mass and radius can be reduced to about the 1% level, it becomes possible to say some interesting things about Sirius B, in particular, and other white dwarfs, in general. For example, what is the mix of carbon and oxygen nuclei in the degenerate interior, how old is the white dwarf, and from what kind of star is it descended? In order to obtain this level of precision, and answer such questions, it has been necessary to make use of a variety of observations from space. These include a new determination of the distance to the Sirius system, and for Sirius B, improved measurements of its brightness, its temperature, its surface gravity, and a very high accuracy estimate of its gravitational redshift. Over the past seven years, these measurements, by my colleagues and myself, have involved the use of four different spacecraft in earth orbit. In turn these space-based observations prompted a revised assessment of the orbit of Sirius and the masses of Sirius A and B.

The most direct way to measure the mass of a pair of binary stars, such as Sirius A and B, is to observe the apparent orbits of both stars. This is the method used for nearly two centuries, since the days of Herschel and Bessel, but it requires time and care. Observing binary stars and determining their orbits was a major theme of 19th century astronomy and it eventually succeeded in establishing the very first estimates of stellar masses, other than that of the sun. In the case of Sirius, these efforts began with the first detection of Sirius B in 1862. During the past 146 years there have been over 2000 independent observations of the angular separation and position angle of the two stars. Most of these were visual observations recorded by astronomers struggling with micrometers at the eyepieces of telescopes and trying to measure the relative positions of two shimmering images, one vastly brighter than the other. In the late 1950s a new, more accurate, photographic technique was systematically employed which gradually overshadowed visual observations. The 120-year long tradition of visual observations of Sirius B finally ceased, with the last recorded observations made in 1982 at a small observatory in the former nation of Yugoslavia. The photographic observations were primarily the efforts of two individuals, G. B. van Albada at Bosscha Observatory on the Indonesian island of Java, and Irving Lindenblad at the U.S. Naval Observatory in Washington, D.C. The observations of Sirius at the U.S. Naval Observatory ceased in 1976, while observations at Bosscha continued until 1986.

There are several motivations for going back to refine the orbit of Sirius. The existing orbital, or dynamical mass, was based on the work of Willem van den Bos and did not include any observations after 1960 and very little of the more accurate photographic data. Thus, there exists the prospect of improving mass estimates by including more recent data. More importantly, my colleagues and I had become involved in a series of *Hubble Space Telescope* (*HST*) observations aimed at imaging Sirius A and B to obtain highly accurate position observations. These *HST* observations, however, would be few in number and would span less than a decade. If we were to get the most out of the *Hubble* observations we would have to also make the optimum use of the long series of historical observations.

Seven parameters are required to completely describe the apparent binary orbit of two stars. The most important of these is the orbital period, which together with the true separation of the two stars, fixes the total mass of the system. Determining the orbital period is a long-term process that necessarily demands observations extending over a substantial fraction of the orbit. In the case of Sirius this means almost 50 years. The commitment of the early visual observers to posterity can be appreciated by the fact that the orbit period of Sirius exceeds the active careers of even the most robust astronomers. Indeed, the first accurate determinations of the mass of Sirius B only became available at the close of the 19th century, nearly a half century after its discovery. Today the primary importance of these early observations principally lies in their ability to constrain estimates of the orbital period, *P*. Observations of successive orbits serve to markedly reduce the uncertainty in the orbital period. Today, with observations of Sirius A and B spanning nearly three complete orbits, this parameter is well determined to within a small uncertainty. The remaining six required parameters are more subtle. These include the exact date, or epoch, T, when Sirius B passed through its apoapse, or closest approach to Sirius A, and the eccentricity, e, or the relative elongation of its elliptical orbit. A fourth parameter is the mean angular separation, a, of the two stars. Together, these four parameters serve to fully define the motion of the Sirius A and B system in the true plane of its orbit.

If we were privileged to be internal observers within the Sirius system, this is all we would need to know to determine the relative orbit of Sirius B with respect to Sirius A and to calculate the subsequent orbit into the past and well into the future. Unfortunately, we are remote external observers and must accept the point of view that nature has provided. We see the orbit only as it appears, projected onto the two-dimensional framework of the sky. This restricted viewpoint necessarily gives a distorted perspective of the true orbit and we are forced to determine the three remaining angular parameters that define the spatial orientation in which we view the orbit. The most critical of these is the apparent inclination, i, or the angle at which we see the orbit plane. The inclination which is defined as 90°, if we see the orbit face-on, as viewed from its north pole, or 0°, if we could view the orbit exactly in its plane, in which case the two stars would mutually eclipse one another twice during each orbit. In most instances, we are at an inclination somewhere in between and we perceive the orbit distorted by perspective, much like a round coin appears elliptical if viewed obliquely. The other two angles define the orientation of the apparent orbit with respect to the east–west reference frame of the sky. The first, called the ascending node, is the point at which Sirius B appears to go from south to north relative to Sirius A in the plane of the sky. The second, called the line of apsides, defines the orientation line of periapse passage with respect to the plane of the sky. Sorting all of these quantities out is a complex task and demands a considerable number of accurate observations over a substantial portion of an orbit.

In 1960 a South African astronomer named Willem van den Bos (1896–1974) sat down with a century's worth of accumulated data and worked out by hand a definitive orbit for Sirius B with respect to Sirius A. Ever since that time the van den Bos orbit has been the gold standard for discussions of the relative orbit of the Sirius system. In preparation for the objective of determining the mass of Sirius B, I collected all of the historical observations that I could find. This involved consulting old compilations and then trying to locate the original publications in which they appeared. For the most difficult and obscure publications this involved trips to the U.S. Naval Observatory in Washington, D.C. and the Harvard College Observatory in Boston. In the end I was able to locate all but a handful of these historical observations, and in the process uncover a number that had not been included in earlier compilations and catalogs. Over 125 observers, including some of the most famous names in observational astronomy, such as Father Angelo Secchi, Wilhelm and Otto Struve, Simon Newcomb, E. E. Bernard, Asaph Hall, and even T. J. J. See, have all contributed large numbers of observations of Sirius B. Fittingly many of these observations were made with the series of large refractors built by the Clarks during the 19th century (Chapter 5). The distribution of these observations on the sky is shown in Figure 12.2

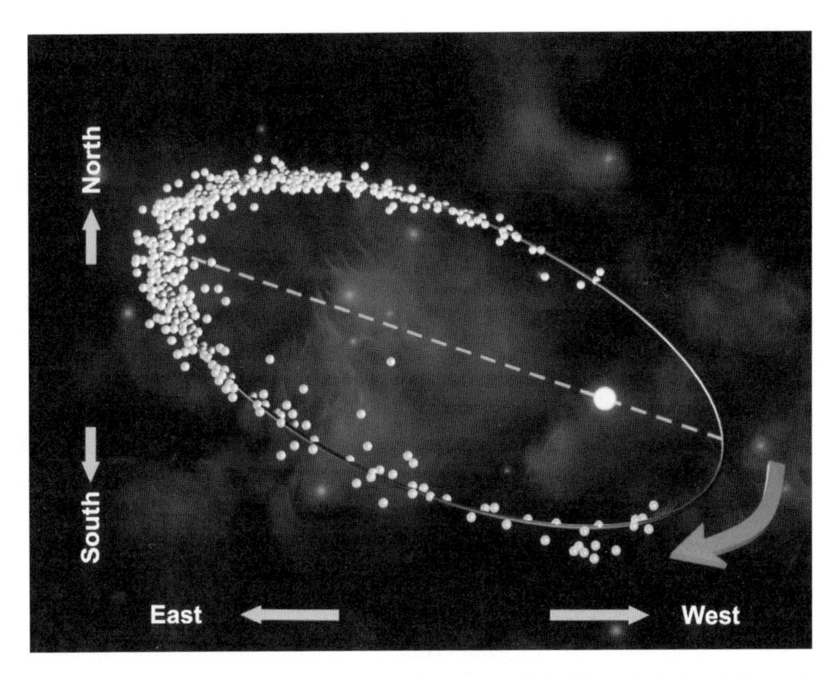

Figure 12.2. The orbit of Sirius B showing the locations of the historical visual (large dots) and photographic observations (small dots).

(see also color section) where the portion of the orbit within 3 arc seconds of Sirius A contains no observations due to the glare of Sirius A. Considered as a function of time, these observations span nearly 140 years and reveal the changing tempo of observations over the last three orbits of Sirius B. The gaps in the record seen in Figure 12.3 (see also color section) correspond to the periods near periapse when Sirius B cannot be seen. The decline in the number of visual observations after the Second World War is largely due to the application of photographic observations beginning in the late 1950s. The only positional observations now being obtained are from our previously mentioned *Hubble Space Telescope* program.

The new orbit includes many observations which postdate the van den Bos orbit of 1960 as well as all of the photographic observations. The improved orbital parameters include the period of 50.075 ± 0.013 years with a most recent periapse passage of 1994.402 ± 0.037 (June 1 ± 3 days, 1994). The semi-major axis and eccentricity are 7.50 arc seconds and 0.59657 ± 0.0013, respectively. The important inclination angle is $138.476 \pm 0.298°$ (the larger than 90° inclination means that we view the orbit from "its south pole", so that Sirius B appears to move counter-clockwise about Sirius A).

The relative orbit of Sirius B about Sirius A is not sufficient by itself to either determine the relative masses of the two stars or their individual masses. Two additional pieces of information are required to achieve these goals. In order to measure the relative masses of the two stars, it is necessary to locate the common center of gravity of the Sirius system, the point in space about which the two stars appear to orbit one another. The relative distances of the two stars from this point provide the ratio of the

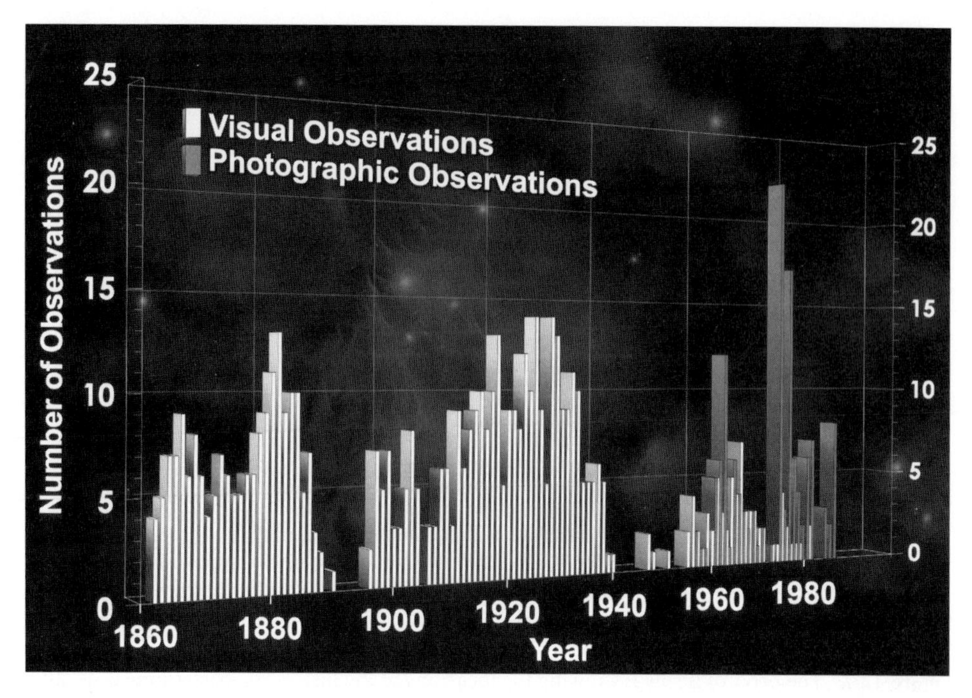

Figure 12.3. Observations of Sirius B as a function of time. The light bars represent the number of visual observations in a given year and the dark are photographic. The gaps near 1896 and 1944 correspond to the period when Sirius B was nearest Sirius and lost in its glare.

masses of the two stars. The 1978 work of George and Carolyn Gatewood, who analyzed the apparent motion of Sirius A about the center of mass of the system, provides this key measurement, establishing that the ratio of the mass of Sirius A to Sirius B is very nearly 2:1.

The final critical piece of information that fully establishes the absolute masses and physical scale of the Sirius system is its distance. Distance turns angular sizes into physical dimensions. The distance to the Sirius system comes directly from measurements of the parallax to Sirius A. These parallax measurements are discussed below, but for the moment they can be anticipated and directly used to give the masses. The new orbit (and parallax) gives the dynamical or gravitational masses of Sirius A and B, which are 2.02 ± 0.03 and 1.00 ± 0.01 solar masses, respectively. As we will see, the present $\sim 1\%$ uncertainties in the dynamical masses of both stars are largely due to the uncertainty in the parallax.

As mentioned in Chapter 6, the first accurate visual measurement of the trigonometric parallax distance to Sirius A was not made until 1898. By the 1970s the star's photographic parallax had been refined to 380 milli-arc seconds. The limitations of ground-based parallax measurement have always been systematic errors imposed by turbulence in the earth's atmosphere. Any significant improvement in the distance to Sirius would have to come from observations from space.

In 1989 the European Space Agency launched a unique satellite called, appropriately enough, *Hipparcos*. This satellite was specifically designed to accurately measure the positions, motions, and parallaxes of over 100,000 stars, providing the most accurate star catalog ever constructed. Operating free of the earth's atmosphere, *Hipparcos* scanned the entire sky for $3\frac{1}{3}$ years, carefully recording the relative angles between stars as they passed through two instrumental fields of view separated by a fixed arc of 58 degrees. When the *Hipparcos* data were finally released to the public in 1997 they revolutionized the study of stellar distances and motions. *Hipparcos* was able to achieve a level of accuracy of 2 milli-arc seconds for trigonometric parallaxes. This effectively means that stellar distances can be measured to within an accuracy of about 10% out to distances of 50 parsecs (160 light years). Sirius A, because of its extreme brightness, and the orbital motion associated with Sirius B, required special attention. Nevertheless, *Hipparcos* measured the parallax of Sirius A to be 379.22 ± 1.58 milli-arc seconds. The *Hipparcos* measurements are significant because they were made almost entirely independently of traditional ground-based observations. First, they are true absolute parallax measurements—that is, no corrections need to be applied for subtle parallax shifts of background stars. Second, they are based on a huge set of data obtained from a single well-characterized instrument. It is very reassuring that the *Hipparcos* parallax and the average of the best ground-based measurements of the trigonometric parallax for Sirius are in good agreement. The weighted mean of the ground-based and *Hipparcos* parallaxes is 380.02 ± 1.28 milli-arc seconds. This corresponds to a distance of 2.8 parsecs, or 8.4 light years, and makes Sirius the sun's fifth nearest stellar neighbor and Sirius B the nearest white dwarf. The relative uncertainty represents about 0.3% of the distance, which is one of the best trigonometrically determined stellar distances. Nevertheless, this relatively small uncertainty in distance continues to be the limiting factor in determining the dynamical masses of Sirius A and B, since orbital mass determinations depend on the third power of the parallax. This amplifies any uncertainty in the parallax by a factor of 3, so that a 0.3% uncertainty in parallax will result in a $\sim$1% uncertainty in mass.

If the minuscule angular shifts associated with stellar distances are difficult to measure directly, then stellar diameters are even more daunting. For example, the parallax of Sirius A is an observable 380 milli-arc seconds, but its angular diameter is some 60 times smaller, about 6 milli-arc seconds. Even in the absence of the earth's atmosphere, it would require a telescope of nearly three meters in diameter to only just resolve such a small disk at optical wavelengths. The 2.4 m diameter *Hubble Space Telescope*, orbiting above the earth's atmosphere, does not quite achieve this resolution. There are, however, several sophisticated methods that can yield direct measurements of the angular diameters of bright stars such as Sirius A. Both methods go by the name of interferometry. The first, which was pioneered by Albert Michelson in the 1920s, relies on the classical wave properties of light to form an interference pattern that can be used to measure a stellar diameter (Chapter 7). The second, which was first employed on Sirius in 1957, uses the quantum properties of light and the fact that photons from the surface of a star are not entirely independent of one another when they arrive at the telescope.

During the unusually good English winter of 1955/56, a young radio astronomer named Robert Hanbury Brown (1916–2002) constructed a most unlikely telescope in the shadow of the giant 250-ft Jodrell Bank Radio Antenna in Cheshire in northwest England. During the Second World War Hanbury Brown had worked as an electrical engineer on the development of British radar. After the war, he returned to the University of Manchester, to obtain his Ph.D. in astronomy, where he applied his wartime electronic skills to studying the heavens with military surplus electronics. Although he was one of the major pioneers in radio astronomy at Jodrell Bank, Hanbury Brown had developed a clever idea that he thought might work to measure the diameters of stars using light and optical telescopes. He and a physicist named Richard Twiss had worked out a way to do "interferometry" without measuring the phases of the light waves. Called intensity interferometry, Hanbury Brown and Twiss showed that the photons of light from a star arrived in pairs. If some way could be found to *precisely* measure the arrival time of these paired photons at telescopes separated by some distance, then a correlation could be observed between the arrival times of photons at the two telescopes. How this correlation would change with the distance separating the telescopes was described by an equation that Hanbury Brown and Twiss worked out. Measuring this correlation, as a function of the separation distance, would thus yield the apparent angular diameter of the star. This was truly a novel idea, one that many physicists at first thought to be nonsense. Nevertheless, Hanbury Brown built a prototype of his unique instrument, fashioning the two telescopes out of searchlight mirrors that had probably seen earlier service during the Battle of Britain. Throughout the months of November 1955 to March 1956, whenever the moon was in its first or last phase and when Sirius was within two hours of the meridian, he collected light from Sirius and electronically combined the output signals from the two telescopes. In order to measure the slight difference in arrival times of the photon pairs at each telescope, Hanbury Brown and Twiss fed the output pulses from each photomultiplier tube through cables of different lengths so that the electronic pulses, traveling near the speed of light, arrived at the electronic correlator simultaneously. In the end it required over 18 hours of observations to measure the correlation at four different separations of the two telescopes, extending out to a distance of 30 feet. Hanbury Brown and Twiss announced their success in *Nature*, reporting that Sirius A had a diameter of 6.8 ± 0.5 milli-arc seconds. The singular problem was that Sirius A was the only star bright enough to be measured with their prototype instrument.

In 1962 Robert Hanbury Brown left the cloudy skies of England to build a much more elaborate version of his intensity interferometer at Narrabri, on the plains of New South Wales, Australia. There, he constructed a larger interferometer to measure the diameters of the brightest stars. Optically the two telescopes were crude "light buckets" designed not to form a stellar image but to concentrate the light onto a photomultiplier tube at the focus of each aperture. They were squat short-focus parabolic mirrors, which resembled radar dishes more than telescope mirrors. Each telescope was mounted on a movable platform that could be wheeled around a circular track 200 m in diameter. The "eyepiece" of the "telescope" was a rack of electronics, with which Hanbury Brown was able to precisely measure the relative arrival times of the detected photons at each photomultiplier tube. As before, stellar diameters were determined from the correlations between arriving photons measured at different telescope separations. Even with the larger Narrabri light collectors, measuring the diameter of a single star required hundreds of hours. By 1976 Hanbury Brown had measured the angular diameters of 32 southern hemisphere stars. For Sirius, he obtained an improved diameter of 5.96 ± 0.07 milli-arc seconds. Unfortunately, Hanbury Brown's technique was still basically inefficient, requiring long observing times in order to accumulate the statistics on the arrival times of hundreds of millions of photons, and consequently worked best only on the very brightest stars. The Narrabri project effectively ended when Hanbury Brown had observed all the bright stars accessible from Australia.

Improvements in optics have led to a revival of the interferometric methods pioneered by Michelson and Pease in the 1920s. Several interferometric systems are currently in operation around the world and have succeeded in measuring the diameters of a number of stars. Measurements at the Very Large Telescope Interferometer located at the European Southern Observatory on Mt. Parnal in Chile and at the Mark III Optical Interferometer instrument on Mt. Wilson, where Michelson and Pease measured the first stellar diameters eighty years earlier, have recently measured the angular diameter of Sirius at 6.039 ± 0.0019 milli-arc seconds and 5.993 ± 0.108 milli-arc seconds, respectively. Using the more precise former value, and the distance to Sirius, directly defines the diameter of Sirius relative to that of the sun: 1.712 ± 0.0025 solar radii. Measuring the radius of Sirius B, which is some 250 times smaller must rely on completely different, indirect techniques.

Although Sirius B is the nearest and brightest of the white dwarfs, it has proved to be one of the most difficult to observe with ground-based telescopes. The difficulty of course lies in the fact that the white dwarf is always bathed in the glare of Sirius A. Even during periods of maximum separation, when the two stars are about 11 arc seconds from each other, making accurate observations of Sirius B has proved a challenge. For example, until very recently astronomers have lacked even a good estimate of how bright it actually is. A consensus of the best existing ground-based estimates gives a visual V-band magnitude of 8.44 ± 0.06, or about a 6% uncertainty. Were Sirius B an isolated white dwarf, we could easily determine its V magnitude to 1% or better. A similar situation existed for the temperature of Sirius B, which over the years has ranged from 32,500 K to the present value of approximately 25,193 K. All of this observational uncertainty in the brightness and temperature directly translates into an embarrassingly large uncertainly on the radius of the star. Our very best determinations of the radius of Sirius B come from space and are based on accurate measurements of the absolute brightness of the star at one or more wavelengths.

The basic method is the one Eddington used when he first estimated the radius of Sirius B (Chapter 8). Three things are required, the distance to the star, its effective temperature, and its absolute brightness at some wavelength. There is also a fourth element involving the detailed description of how the star emits its radiation as a function of wavelength. This model of the star's atmosphere basically predicts the flux of radiation (light) from each square centimeter per second at the surface of the star.

It relies on a detailed mathematical description of how radiation flows or, more realistically, percolates up from below through the atmosphere of the star and into space. On its journey from the degenerate core of the star to the surface, the energy travels as photons which are absorbed and reemitted countless numbers of times by atoms in the outer layers of the star. It may seem like a hopelessly complicated process to follow in detail, but the simple expedient of taking averages of the outcomes of many absorptions and emissions turns out to be a highly accurate means of determining the overall outcomes. In practice, the stellar atmosphere is sliced into several hundred imaginary vertically stacked layers, each characterized by a specific temperature, pressure, and ionization state. A computer program then uses the laws of physics to determine, on average, how much radiation at each wavelength enters at the bottom of each layer and how much leaves from the top of the layer.

The fortunate thing about white dwarfs like Sirius B is that the details of the physics are particularly simple because the atmosphere consists almost exclusively of nature's simplest atom, hydrogen, and its components, the proton and the electron. The descriptions of how these components interact with light and one with another are relatively simple and well known to astronomers and physicists. This is not the case with many heavier elements where the number of possible interactions increases dramatically and approximation techniques are frequently necessary to describe the outcomes. For stars such as Sirius B it is possible to compute a detailed spectrum of the light given off by the star as a function of just two variables, the star's effective temperature and its surface gravity. The effective temperature is a single defined number, which is used to characterize the entire range of temperatures in a stellar atmosphere. It is determined by the temperature of a black body that has the same total energy flux as the stellar atmosphere. The effective temperature can also be thought of as approximately the average temperature of the layer of the atmosphere where the bulk of the light from the star originates. The surface gravity is the gravitational acceleration at the star's surface and it determines the pressure in each layer in the stellar atmosphere. The surface gravity of a star (or planet) depends on its mass divided by the square of its radius. In units of centimeters per second squared the surface gravity g of the earth is 980, the sun is 15,800, and that of Sirius B is 468,000,000. In stars, these quantities are usually stated as logarithms so the gravity of Sirius B, or its $\log g$, is said to be 8.67.

Once the temperature and gravity are specified, the spectrum of a white dwarf like Sirius B with a pure hydrogen atmosphere is completely defined at all wavelengths. The confidence in these calculations is such that the stated uncertainties are presently on the order of just 1%. This is not generally true of most other stars, where a detailed knowledge of the relative abundances of dozens of other elements in the atmosphere must be known, or at least approximated. Once the spectrum is known, the determination of its radius is a matter of simple geometry. The flux at the surface of a star can be related to its flux observed at the earth (after correcting for any absorption of the light due to interstellar dust and gas and/or the earth's atmosphere) by a simple ratio, the square of the stellar radius divided by the square of its distance from the sun. This is a small number that is effectively the square of the angular diameter of a star.

Our very best determination of the radius of Sirius B now comes from the *Hubble Space Telescope*. The observations, which are described in more detail later, provide a number of fundamental details about Sirius B. One of these is an absolute measurement of the flux of light from the star at optical wavelengths, which is free from contamination from Sirius A. From this flux it is possible to determine for the first time an accurate V magnitude of Sirius B, which turns out to be 8.48 ± 0.01, close to the older ground-based estimates but with a much reduced uncertainty. From this value the radius of Sirius B can be computed in solar radii as 0.0084 ± 0.0001, this is about 5840 km, or 92% the radius of the earth.

Fortunately new observations of Sirius B from space have finally overcome most of these difficulties associated with the light from Sirius A, so that we presently know the mass and radius of Sirius B to a better precision than we do for any other white dwarf. In the last six years Sirius B has been successfully observed by no fewer than four spacecraft. These include *Hipparcos*, the *Extreme Ultraviolet Explorer* (*EUVE*), the *Hubble Space Telescope* (*HST*), the *Far Ultraviolet Spectroscopic Explorer* (*FUSE*); each contributed critical observational data.

In the mid-1970s, Sirius B had one remaining lesson to teach space-age astronomers. A small Dutch satellite, *ANS*, or the *Astronomical Netherlands Satellite*, detected X-rays coming from Sirius. *ANS* was primarily a mission to explore the sky at ultraviolet wavelengths but it also carried one of first soft (low-energy, longer wavelength) X-ray telescopes to orbit the earth. In a very brief letter to the journal *Nature* in August of 1975, the Dutch astronomers established that the X-ray telescope had detected soft X-rays from Sirius, in the instrument's sensitive band between 44 Å and 60 Å, but they were at a loss to identify whether the X-rays came from Sirius A or Sirius B. By December, in a second paper, they had settled on Sirius B as the likely source of the X-rays and suggested that the X-rays might be generated by a corona surrounding the star. An alternative explanation involving X-rays from the warm photosphere itself was rejected by a number of astronomers over the next several years. The reason for rejecting the photosphere was that the effective temperature of Sirius B was simply too cool to produce the copious amounts of observed soft X-rays. On the other hand, astronomers were quite familiar with stellar coronae, especially that of the sun. The sun's corona is the broad whitish halo of radiance that seems to blaze out from the darkened solar disk during a total solar eclipse. The coronal light comes from a rarefied but very hot and extensive halo of gas surrounding the sun. At a temperature of approximately 2 million degrees K, the corona is responsible for much of the sun's X-ray output. Stars similar to the sun also have coronae and several had been observed by the early X-ray satellites. If somehow Sirius B possessed a corona then this might well explain the observations. The problem was that it was difficult to envision a corona around a white dwarf. In the case of the sun, the corona derives its heat from the maelstrom of convective and magnetic activity expressed at the solar surface. In a white dwarf such as Sirius B, the atmosphere is quite different. There are no convective atmospheric motions and no sensible magnetic field. The radiant energy flows smoothly out of such an atmosphere, which, although quite dense and hot, is very stable and placid. There seemed to be no way to heat a corona, or to maintain one if it ever managed to form. Nevertheless, coronas remained an attractive idea, and theoretical attempts were made to use them to explain the X-rays from Sirius B.

The resolution of the dilemma of how white dwarfs manage to produce X-rays finally came in 1976, from Professor Harry Shipman, at the University of Delaware. The answer turned out to be relatively simple. Hot pure hydrogen white dwarf atmospheres, as Shipman pointed out, become increasingly transparent to the photons of shorter wavelength X-rays. Effectively, astronomers were viewing much deeper and hotter layers of the star with X-rays, than they were in the optical. X-rays, originating from layers several times hotter than the optical photosphere, were streaming out relatively unabsorbed, so that when observing at shorter wavelengths one was sampling deeper much hotter layers of the stellar atmosphere. With this realization, and including the proper stellar opacities, models of the atmospheres of Sirius B and other white dwarfs could easily be made to reproduce what was observed. A corollary to Shipman's explanation was that it only worked well for white dwarfs whose atmospheres were composed almost exclusively of pure hydrogen. Add just a bit of helium or other heavier elements, even at a few parts per ten thousand, and the atmosphere becomes increasingly opaque, and finally chokes off the X-ray emission altogether.

The increasing transparency of hydrogen to short-wavelength X-rays has an important and opposite consequence at longer wavelengths, where the ability of hydrogen to absorb X-rays increases rapidly. From the longest wavelength X-rays at 100 Å up to wavelengths of about 900 Å, hydrogen gas becomes increasingly opaque. This wavelength range is traditionally called the extreme ultraviolet (EUV). Interstellar space is filled with small amounts of atomic hydrogen and, as a consequence, EUV radiation tends to be rapidly absorbed before traveling very far. This was a well-known fact in the 1950s, and astronomers had calculated that if one took the average density of interstellar hydrogen, of 1 atom per cubic centimeter, then only a few of the very closest stars might be dimly observed in this band. Therefore, at a time when astronomers were opening new windows on the universe in the ultraviolet, X-rays, and gamma rays, and in the infrared and radio bands it seemed as if EUV astronomy would only be practical for objects within the solar system, or at best the very nearest stars.

Not everyone accepted this pessimistic verdict. One in particular was Professor Stuart Bowyer of the University of California at Berkeley, who persistently took a contrary view. He argued that while hydrogen was indeed a serious source of opacity, its distribution in interstellar space was far from uniform and that certain directions might contain far less than the frequently quoted average value. Moreover, portions of the interstellar medium might also be highly ionized and thus be even more transparent at EUV wavelengths. Bowyer (who was the author's thesis advisor) built and flew crude EUV detectors on sounding rockets in the early 1970s and even proposed satellite instruments to NASA to conduct observations in the EUV. His efforts finally paid off in 1976, when NASA flew one of his instruments on the *Apollo–Soyuz* mission. *Apollo–Soyuz* was a milestone in the cooperation between the American and Soviet space programs, and involved the orbital linkup of an American *Apollo* capsule and a Soviet *Soyuz* space module. In an effort to include some science along with the international space diplomacy, NASA added Bowyer's EUV telescope to the mission. The telescope consisted of a set of nested metallic mirrors which crudely focused EUV radiation onto a set of detectors, or channels, each covered by a different thin-film filter that provided each channel a different wavelength band, from 100 to 300 Å. During the 8-day mission, the *Apollo–Soyuz* telescope detected six EUV sources, including two white dwarfs. *Apollo–Soyuz* did look at Sirius B, but in hindsight it turned out not to have been sensitive enough to detect it. The detection of two other white dwarfs, however, lent further weight to Shipman's demonstration that white dwarfs were capable of producing soft X-rays and extreme UV radiation well in excess of what a simple consideration of the photospheric temperatures would seem to permit.

The EUV window on the universe had been forced open, but it proved a challenge to keep it from closing. Bowyer and his Berkeley team proposed several dedicated and more ambitious satellite experiments to follow up on the *Apollo–Soyuz* discoveries, but it would be nearly fifteen years before another EUV mission was flown, and it would be the British with their Wide Field Camera on the *Roentgen Satellite* (*ROSAT*) mission launched in 1990, and not Berkeley, who got there first. NASA initiated and then abandoned several EUV missions and it was only Bowyer's dogged determination that succeeded in the 1992 launch of the *Extreme Ultraviolet Explorer* (*EUVE*) satellite. *EUVE* was a NASA mission to study the sky in three extreme ultraviolet (EUV) wavelength bands between 80 Å and 760 Å. The satellite's prime mission was to photometrically map the entire sky at these wavelengths subsequently uncovering 700 sources of EUV radiation. Among these sources was Sirius, which was detected by both the *ROSAT* Wide Field Camera and by *EUVE*.

In addition to the EUV photometric telescope, *EUVE* also had a set of three spectrometers designed to obtain spectra of sources at EUV wavelengths. The satellite obtained the first EUV spectra of Sirius B in a period between November 27 and December 2, 1996. Now for the first time it was possible to clearly determine where the flux was coming from. As Harry Shipman had pointed out twenty years earlier, the flux from Sirius B was emitted from deep within a hot photosphere of pure hydrogen. A nearly featureless spectrum emerged from the noise at about 200 Å, peaked near 250 Å, and decreased to invisibility again at 350 Å (Figure 12.4). The only features present were weak lines due to a small amount of interstellar helium. Because of the extreme sensitivity of the emitted spectrum to temperature it was possible to determine, with great precision, the effective temperature of Sirius B: $24{,}790 \pm 100$ K. The spectrum was also shaped by absorption from interstellar gas, primarily by neutral atomic hydrogen, but also helium. In fact the downturn in the flux near 350 Å was primarily due to interstellar absorption. Although the opacity of the interstellar medium was high at these wavelengths the amount of gas turned out to be quite small: 1.0×10^{18} atoms per square centimeter, spread out over the entire distance to Sirius. The interstellar gas that absorbs the EUV light from Sirius B corresponds to an average density of only about one neutral hydrogen atom per cubic centimeter. By comparison the air in a room contains about 10^{19} molecules per cubic centimeter.

The results on Sirius from *EUVE* highlight the enormous advantage of observing the white dwarf from space. In the last few years, efforts to improve the estimates of the temperature, radius, and mass of Sirius B from space have made it possible to

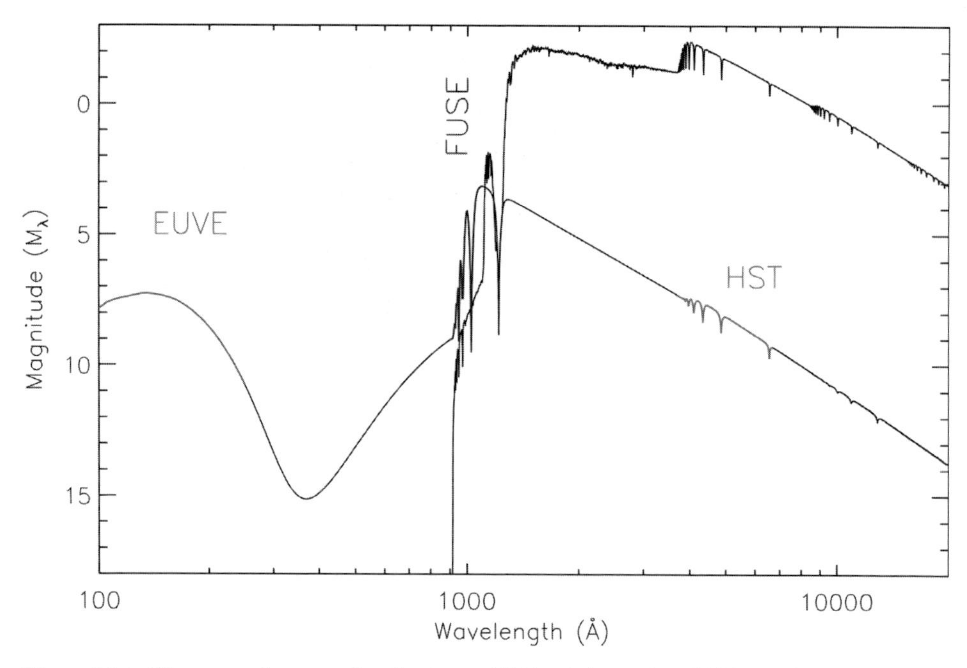

Figure 12.4. The spectral energy distributions of Sirius A (upper) and Sirius B (lower). The flux from Sirius B matches and exceeds that of Sirius A in the region near 1000 Å; below 900 Å the flux from Sirius A falls abruptly. The wavelength regions where various spacecraft have observed Sirius B are indicated.

determine these quantities with unprecedented accuracy. The ultimate goal of these space-based observations is to make a critical comparison between the primary dynamical mass estimate of Sirius B and corresponding estimates indirectly derived from spectroscopy. Dynamical or orbital mass determinations are generally regarded as the best way to determine stellar masses and are completely independent of spectroscopic methods. As such they are a standard against which astronomers can test their theories which describe the stellar atmosphere of white dwarfs, their degenerate interiors, and the thermal evolution of these stars.

The *Far Ultraviolet Spectroscopic Explorer* (*FUSE*) was designed to conduct detailed observations of the narrow wavelength band between 900 Å and 1180 Å, called the Far Ultraviolet (see Figure 12.4). This band is bounded on the short-wavelength side by the opaque curtain of interstellar hydrogen and on the long-wavelength side by the intense background of scattered light emitted by atomic hydrogen surrounding the earth (Chapter 13). Within this narrow wavelength range, nature has crowded the important spectral lines of many molecules and elements. By the early 1990s there was a scientific eagerness to explore this region beyond the reach of the *Hubble Space Telescope*. The *FUSE* spacecraft was launched by NASA in 1997 and contains a bank of eight spectroscopic channels designed to make sensitive high-spectral resolution observations of stars, galaxies, and interstellar gas.

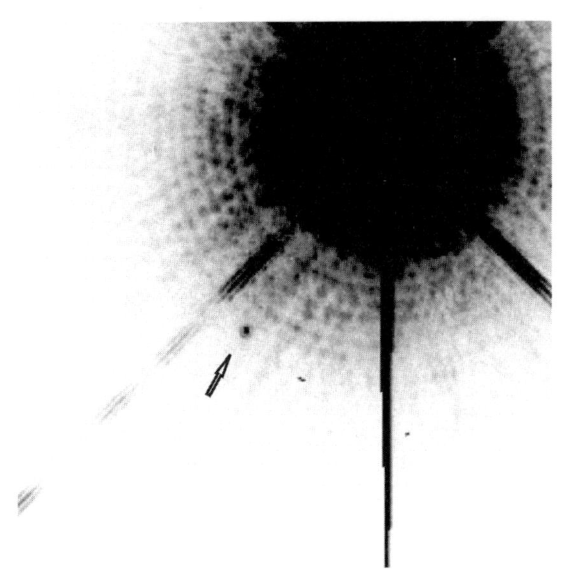

Figure 12.5. A *Hubble Space Telescope* Wide Field Camera image of Sirius A and B showing the white dwarf (arrow). The bright diagonal spikes are diffracted light from the telescope secondary mirror support structure. The bright vertical bar is an overexposure feature in the detector and is not light that will enter the slit of the *Space Telescope Imaging Spectrograph (STIS)* instrument. The entire spacecraft was rotated so the image of Sirius B fell midway between the diffraction spikes (from Barstow *et al.*, 2005).

In 2001 the author led a team of astronomers who proposed to get the first good look at Sirius B in the Far UV since the earlier *Voyager* observations nearly two decades before. These observations were not attempted early in the *FUSE* mission because it was necessary to wait for Sirius B to move far enough from Sirius A to make such observations possible. By late 2001 the two stars had separated to a distance of 5.4 arc seconds and it was decided that a try could be made. Much careful planning had to go into the observations since any error could accidentally place Sirius A in the instrumental aperture. This would not only doom the observation but possibly permanently damage the sensitive *FUSE* detectors. With the help of Dr. Jeff Kruk at the Johns Hopkins University, the institution that built the *FUSE* instruments and operates the *FUSE* mission, we rolled the spacecraft so that the 10×30 arc second rectangular aperture of the spectrometers was centered on Sirius B and oriented perpendicular to the line joining the two stars. In this way Sirius A was just excluded, while Sirius B was fully within the slit. It was also necessary to turn off the instrument channels covering the longest wavelengths so that the light from Sirius B would not exceed the brightness limit of the detectors.

Sirius B was successfully observed by *FUSE* for several hours on 25 November 2001. The observations revealed the spectrum of the white dwarf stretching from 900 Å to 1100 Å and indicated, to our great delight, that the data were virtually free from scattered light from Sirius A. Our goals in obtaining these observations were

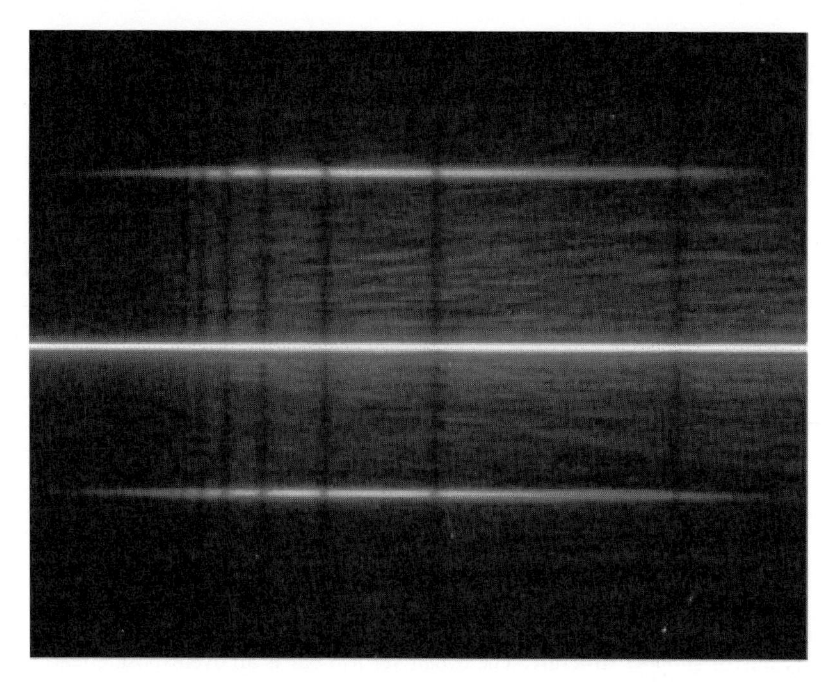

Figure 12.6. The *STIS* spectrum of Sirius B (the bright central line). The scattered light from the diffraction spikes of Sirius A appear above and below the Sirius B spectrum. A low level of scattered light from the primary image of Sirius A is seen as a diffuse background. The faint dark vertical lines cutting the spectra are the hydrogen lines from Sirius A (from Barstow *et al.*, 2005).

threefold. First, we wanted to use the Lyman lines (Appendix A) from hydrogen, which extend from 912 Å to 1216 Å to determine the gravitational redshift of Sirius B. Second, we wanted to measure its surface gravity. Third, we wanted to determine how bright the star was at these wavelengths. All these measurements had a single overriding objective: to accurately determine the mass and radius of the white dwarf.

The *Hubble Space Telescope* is the ideal instrument with which to determine the gravitational redshift of Sirius B. The observations, however, had to wait for a number of things to fall into place. When Hubble was launched in 1992, Sirius B had been moving closer to Sirius A, making observations difficult. In addition, the original compliment of Hubble instruments lacked a high-resolution spectrograph that was well suited to make the observation. In 1999, during the third Hubble-servicing mission, the original high-resolution spectrometer was replaced with a more capable instrument, the *Space Telescope Imaging Spectrograph*, or *STIS* (Figure 12.6). This is an extremely powerful instrument capable of operating in dozens of different modes both in the ultraviolet and the optical. Although *STIS* can produce images with its small field of view, its real value comes from the exquisite spectra that it produces. By the time *STIS* was in operation Sirius B was once again moving away from Sirius A, and by 2003 it had reached the point that scattered light was no longer a major concern.

A group of astronomers led by Professor Martin Barstow of the University of Leicester in the United Kingdom, and including myself and Dr. Howard Bond of the Space Telescope Science Institute and others, decided that the time was ripe to propose to use *STIS* to make a precise measurement of the gravitational redshift of Sirius B. The process of submitting such a proposal is in theory relatively straightforward but actually involves a great deal of background preparations. In a standard *HST* proposal just three pages are allotted in which to explain what you want to observe, how you plan to make the observations, and what you plan to do with the observations. Most important of all, however, you must explain to a skeptical review panel, many of whom do not work in your field, why your observations should be given precedence over hundreds of others that have been submitted. In our case the fact that we were observing Sirius proved a double-edged sword. In our favor was the fact that we were proposing to repeat an historic observation on an object familiar to everyone. On the other hand, we had to explain why we were asking to use one of the world's most powerful, expensive, and oversubscribed telescopes to observe the brightest star in the sky. One question review panels often ask is: Couldn't this observation be done from the ground? The answer to that question was easily dealt with: it had been done from the ground, almost thirty years earlier under very favorable conditions with the largest telescope available at the time, and the observations left much to be desired. *Hubble* could clearly do much better.

Hubble observations are handed out in units called orbits, and Hubble orbits the earth eighteen times a day. We were asking for only one 90-minute orbit while some proposals requested hundreds of orbits. There are only so many orbits in a year and each is considered precious, since usually only one observation can be done in a given orbit. *Hubble* review panels are seldom of one view and there are always those who feel the time would be better spent imaging one more galaxy or obtaining the redshift for another quasar. In the end we were granted our single orbit. We had promised in the proposal that with that single orbit we could resolve much of the uncertainty regarding Sirius B and would be able to measure the radius and mass of the star to about 1%.

Once a proposal is accepted, a process of detailed planning begins. On the observer's side this means specifying such things as the precise pointing of the telescope, the exposure times, and grating settings, etc. For those in charge of the spacecraft, these inputs are converted into an exact set of commands to point the telescope, acquire the target, and direct the instrument to take and record the observations. Observations are all placed in queue and are executed automatically at the appointed time. One thing we were particularly concerned with was scattered light from Sirius A. Most of this scattering is contained in four diffraction spikes produced by the telescope's secondary mirror's four support vanes (Figure 12.5). These diffraction spikes, although intrinsic to every stellar image formed by the telescope, are usually too faint to be seen but in a star as bright as Sirius they dominate the image. Calculations showed that Sirius B would be uncomfortably near one of these spikes and so it was decided to roll the spacecraft about its optical axis so that Sirius B would land midway between the spikes in Figure 12.5. The resulting dispersed image is shown in Figure 12.6.

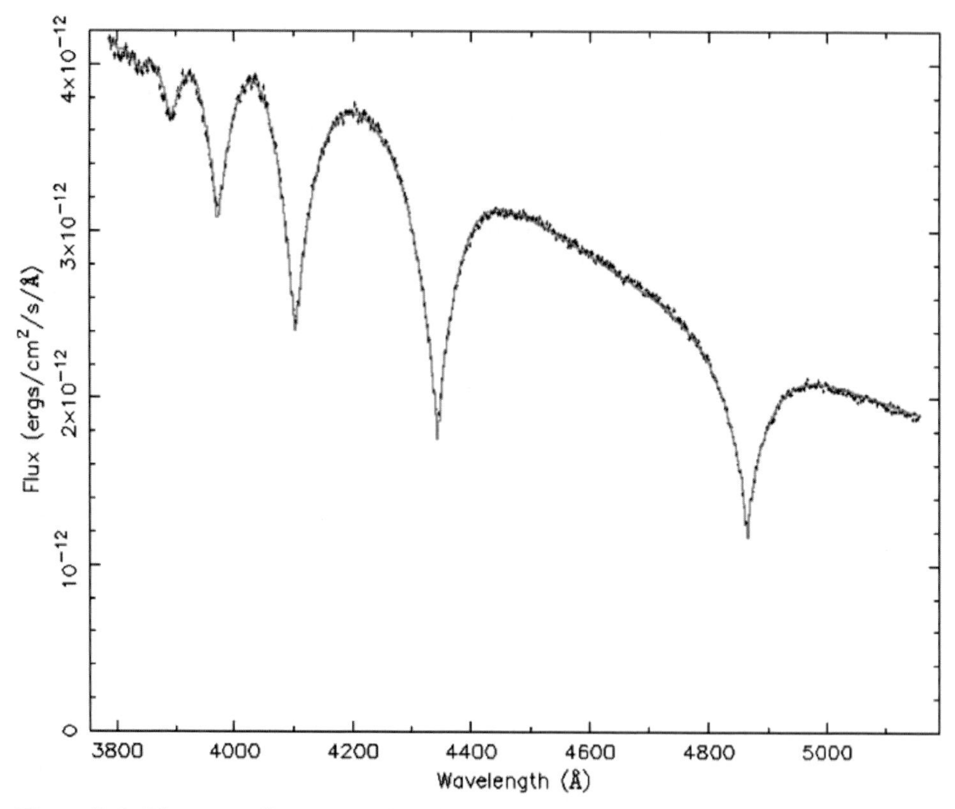

Figure 12.7. The energy flux vs. wavelength spectrum of Sirius B showing its broad hydrogen lines. The solid line is a mathematical model of the Sirius B spectrum from which it is possible to determine the effective temperature and surface gravity of Sirius B (see Appendix B) (from Barstow *et al.*, 2005).

Our observations began at 1:56 p.m. Greenwich Mean Time on February 6, 2004. Hubble had just completed some calibration images with the Advanced Survey Camera and, while still on the daylight side of the earth, Hubble slewed to a point near the earth's limb where Sirius would soon appear. At the same time *STIS* was configured to perform our observations. As the spacecraft was entering the earth's shadow, somewhere over the Indian Ocean, its Fine Guidance System acquired a preselected set of guide stars and made fine-pointing adjustments to place Sirius B in the large aperture of the *STIS* spectrometer, this was followed by a brief "peak-up" exposure to exactly center the star in the aperture. These procedures required about 8 minutes out of the approximately 50 minutes of nighttime observing available during each orbit. Our first exposure was made with the G430L low-dispersion grating that provided a spectrum that extended from 3900 Å to 5000 Å. This gave good spectral coverage of most of the star's hydrogen Balmer lines, which extend from 3650 Å to 6564 Å. We would use these data to determine the brightness, temperature, and gravity of Sirius B. Since for *STIS*, Sirius B is a relatively bright target and required only a 10.5-second exposure, we made three exposures at this grating setting. *STIS* then changed to its G750M medium-resolution grating and tilted it to center the spectrum on the broad hydrogen Balmer alpha line of Sirius B at 6564 Å. This is the observation that would measure the gravitational redshift. After a 90-second exposure it was all over. During our 50 minutes of orbital nighttime we had collected just over 2 minutes of data on Sirius B: the rest was overhead devoted to setting up and configuring the instruments. It should be mentioned that we were extremely fortunate. Just months later, the *STIS* instrument became completely inoperable after the failure of an internal power supply. If *STIS* is ever to operate again, it will have to be repaired by astronauts during a planned Space Shuttle servicing mission.

Ten days after we had observed Sirius B, I was visiting Martin Barstow on the occasion of his elevation to Professor at the University of Leicester. Late one afternoon in his office, after we had discussed other matters, he asked if I would like to see the Sirius B spectra? He had electronically downloaded them several days earlier after they had been reduced, verified, and archived by the Space Telescope Science Institute, in Baltimore. As I looked over his shoulder, he displayed the first set of low-resolution spectra containing the higher order Balmer lines. They looked almost exactly like the theoretical models we had used to plan the observations. In particular, there was no evidence of scattered light which might make the planned analysis difficult or even impossible. A quick fit of the model fluxes to the data would tell us exactly how good they were. Martin ran the fits and overplotted the best fitting model on top of the data. The fit was nearly perfect (see Figure 12.7). More importantly the temperature and gravity for the best-fitting model was very close to the results we had obtained in 1998 with *EUVE*. We were clearly in the right range and there appeared to be no problems with the observation. We next looked at the hydrogen Balmer alpha line spectrum. Again it looked like a textbook plot: in particular the cusp-like center of the line was sharp and well defined, with no evidence of infilling from scattered light. A quick fit to this line confirmed the temperature and gravity were consistent with the results we obtained from the higher order Balmer lines. No one had ever obtained data of this quality on any white dwarf, let alone Sirius B. We made no attempt to measure the redshift that afternoon but I recall that on the flight back to Tucson it seemed as if we had succeeded in getting the observations we had promised in our proposal. However, only several months of work with the data and studying how well they agreed in detail with all our other data would tell for sure. Up to this point things had gone almost perfectly, contrary to almost all other telescope data where it is a constant struggle to remove artifacts, correct for distortions, and improve the statistical uncertainties. It is natural in such circumstances to harbor suspicions in the back of your mind that things were going too well. Martin also had this feeling.

A few weeks after my visit, Martin sent an email with the result of the gravitational redshift. I was expecting something in the range of 70 to 90 km/s, which I was getting from the *FUSE* data and which broadly agreed with the older Mt. Palomar data published in 1971. The number Martin was getting was ten times larger than this. Something was wrong. We discussed where the problem might lie: had the observation been somehow compromised, had the Space Telescope Science Institute applied the wrong wavelength calibration files in reducing the data? The answer came about a month later in another email from Martin. He had discovered that the wavelength scales of the data and the models differed almost exactly by the amount contributed by atmospheric refraction. What did the earth's atmosphere have to do with Hubble observations in space? Traditionally, precise wavelengths of spectroscopic lines are expressed in a laboratory wavelength scale, which contains a small shift due to the reduced speed of light in the earth's atmosphere. For wavelengths below 2000 Å, the vacuum wavelength scale lacking this shift is used. We had incorrectly assumed that Hubble data were specified in vacuum wavelengths. Once the models and the observational data were placed on the same scale the results changed dramatically, in the right direction. The final answer was $80.42 \pm 4.82\,\mathrm{km\,s^{-1}}$.

All of this occurred just one week before Martin was scheduled to make the first presentation of our results on Sirius B to our colleagues at the 14th European White Dwarf Workshop in Kiel, Germany. When Martin delivered his 20-minute talk, he began with a review of the milestones in the study of Sirius B since Bessel first discovered the dark companion in 1844. He was saving the best till last. The final "PowerPoint" plot showed our previous 1998 attempt at determining the mass and radius of Sirius B and how it compared with the mass–radius relation. That result was dominated by a set of long narrow-sloping ellipses which expressed the joint uncertainty in both the mass and the radius. The ellipses passed thought the theoretical mass–radius line but they also passed through much territory that was incompatible with the theoretical mass–radius line. At the center, Martin had plotted our new uncertainty ellipses in red. They were clearly small but it was difficult to see how well they agreed with theory. He then dramatically expanded the inner region of the plot and it was apparent that the theoretical mass–radius relation passed through the center of our uncertainty ellipses at the point expected for a 1.0 solar mass white dwarf. After the talk there were a few questions and technical comments from the audience. Someone wanted to know if thick or thin hydrogen atmosphere layers were used in the models, Martin answered thick. The most memorable reaction came from Harry Shipman, who as a graduate student at Caltech in 1971 had made the historic Palomar measurements from the long thin glass photographic plate spectra of Sirius B obtained at the Palomar 200-inch telescope. Harry, who is seldom at a loss for words, could only comment, "Wow!"

13

Past, Present, and Future

"Ice is the silent language of the peak;
And fire the silent language of the star"
—Conrad Aiken, *And in the Human Heart* (1940)

Having recounted the history of Sirius in human terms it is natural to inquire into its physical and evolutionary history as a system of two stars. At the same time it is also revealing to look into the future and view the fate of the Sirian system and contrast this with the fate of our own sun and solar system. When considering the future of Sirius and the sun, all time scales of any importance are very long, compared with human history. Nevertheless, the important factors that govern the significant events are very similar and fairly well established in both cases. For the evolution of both Sirius and the sun, most of the margins of uncertainty are relatively small. In essence, there is really little room for dramatic unforeseen changes of the type represented by the idea of a "Red Sirius". It is much easier for us to predict the fate of the stars than to use the stars to predict the fate of our lives or that of our civilization.

On the shortest time scales, those covering a few million years, our earthly view of Sirius and the rest of the night sky will undergo dramatic changes. Sirius will not forever remain the brightest star in the sky. It owes its current prominence to its close proximity to the sun. Over the next 60,000 years, the relative motion of Sirius and the sun will cause their mutual distance to decrease. This motion is modest and will lead to fairly modest changes in the apparent brightness of Sirius, but will produce quite dramatic changes in its position in the night sky. Presently Sirius is obliquely approaching the solar system at a somewhat leisurely rate of 8.6 km/s. If we project this three-dimensional motion of Sirius into the future, the star will make its closest approach in about 67,000 years. At that point Sirius will have closed, from its present distance of 8.5 light years, to within 7.6 light years of the sun. As a consequence, its brightness will have increased in magnitude from the present −1.41 to about −1.6.

This represents only a modest 25% increase in apparent brightness and, given the logarithmic response of the human eye, will not result in any dramatic change in its visual appearance; but Sirius will outshine Jupiter and come to rival Venus when that planet is at its brightest. After closest approach, Sirius will begin to recede from the sun and slowly decline in brightness, falling back to its current levels 130,000 years from now. Looking even further into the future, Sirius will continue to fade and by 300,000 years it will have become a 0th magnitude star like Vega. By 3 million years from now it will have assumed the relative obscurity enjoyed by the host of other 5th magnitude stars, which today attract little attention, just as it did in the equally distant past, when Sirius went unnoticed by our Australopithecine ancestors.

The location of Sirius in the sky will change even more dramatically. It is now progressing to the southwest at a rate of one degree every three thousand years. Ten thousand years from now this motion, together with the earth's precession, will make Sirius an exclusive jewel of the southern hemisphere skies, no longer visible in Europe or North America. At this time it will be prominent in the skies of South America, Australia, and Africa. On the same time scale, however, the proper motions of other bright stars will also have changed their relative positions to the extent that many constellations will have noticeably altered their familiar appearance. It is poignant to speculate that in this future time, ten millennia from now, there may remain very few places on the earth—perhaps only in the central regions of the vast southern oceans—where the artificial lights of human population centers will not dominate the night sky to such an extent that just Sirius and a few other bright stars will be the only stars visible to much of humanity.

Looking backwards now, over a much longer time scale, two hundred and thirty eight million years ago (give or take 13 million years): a collapsing cloud of gas and dust gave birth to a host of stars. Many of these were bright blue stars, which have now been dispersed across the sky. The exact present day identity of some of these stars is in some dispute, but they are thought to include many of the brighter stars seen in our night sky. One proposed identification for this family of stars is the "Ursa Major group". As mentioned in Chapter 7, Sirius was first suggested as a member of this group by Ejnar Hertzsprung in 1909, based on its proper motion, and hence it is sometimes also called the "Sirius supercluster". However, as we shall see, recent evidence suggests that the age of Sirius is only half that of the Ursa Major cluster, making Sirius' membership questionable.

At the time Sirius was first formed, the earth was in the midst of the Triassic Period, which witnessed the rise of the dinosaurs. The newly born Sirius, however, appeared very differently than it does today. Two bright blue stars regularly orbited each other in an elongated elliptical orbit every 9.1 years. The brighter star was a huge blue star of spectral type between B6V to B7V, five times the mass of the sun and some 630 times more luminous. The fainter companion, orbiting 8.4 astronomical units away, was today's Sirius A, an A0V star, with twice the mass of the sun. The bright star, which would have resembled Regulus, the brightest star in the constellation of Leo, was furiously burning its hydrogen fuel at a prodigious rate in a series of complex nuclear reactions that results in the net production of helium in its core. It had to do this in order to balance the enormous internal pressure of its mass and to prevent its collapse. In the process the star produced some 1260 times the light of the sun. Because of its greater mass, the larger star was consuming itself at a much higher rate than its less massive companion. After approximately 100 million years the bright star inevitably reached an energy crisis. The hydrogen in its core neared exhaustion and to meet the crushing gravitational demands of its massive outer layers, the star began to burn hydrogen to helium in a shell surrounding the now inert helium core. Although the net energy remained roughly constant, the appearance of the star changed dramatically. In the space of a few million years it went from a modestly sized blue star to a large red star as its outer layers expanded. The star had metamorphosed itself into a huge sullen red giant that now outshone its companion (Sirius A) by nearly fifty to one. A distant extraterrestrial "Ptolemy" would now have observed Sirius B as a single bright red star, the companion being completely unnoticeable.

Shell burning only deferred the crisis. As this source of energy flagged, the star's core contracted and grew in density until it reached the temperature of 100 million degrees K. At this point the star turned to the lucrative but profligate process of burning helium into carbon and oxygen. This state of affairs was not sustainable: the bright star faced increased gravitational demands from its heavy outer layers. Its accumulated capital of helium in the core now became exhausted and a renewed energy crisis ensued. The iron laws of stellar physics would not permit any further deferment of the accumulated energy debt, and the star was not massive enough to undergo any further cycles of nuclear refinancing. A stellar bankruptcy was on the horizon and a fundamental internal reorganization would be necessary.

It happened quickly. After approximately a hundred million years of profitable stellar evolution, in a few hundred thousand years, the star was forced to liquidate its outer layers in a stellar wind that dispersed nearly 4/5ths of its mass into interstellar space. The core, relieved of the heavy burden of its outer layers, now lacked the energy to expand. The nuclear fires were largely extinguished and its energy output had to be steeply scaled back. What remained of the star shrank dramatically and the gravitational force again dramatically increased. To compensate for the intense gravitational pressure, the matter in the interior core of the star underwent a fundamental conversion. The familiar atoms of carbon and oxygen were squeezed out of existence and all that remained was a fluid composed of free electrons and bare stripped atomic nuclei. This fluid no longer adhered to the familiar ideal gas laws that governed the ordinary gaseous matter in normal stars. Rather, it obeyed the quantum mechanical laws of a degenerate gas, established by Fermi and Dirac in 1926. The degenerate electrons now supplied the pressure needed to support the star and its final radius was determined by the theory developed by Chandrasekhar (Chapter 8): a star had died and been resurrected as a white dwarf.

The time is now advanced to about 124 million years ago. The earth is ruled by the fierce carnivorous dinosaurs of the Cretaceous Period. For the Sirius system the roles of the two stars are now completely reversed. Sirius B is now half as massive as Sirius A and almost completely hidden in the glare of its former junior partner. After Sirius B shed most of its mass into interstellar space, the orbit expanded to nearly 20 AU with a more leisurely period of 50 years. The newly formed white dwarf was now also committed to eternally paying back its accumulated debt of gravitational energy by radiating light into space. As it radiated, it cooled, at first it was an extremely hot, tiny point of light with a surface temperature of perhaps 120,000 K. At this temperature the star produced copious quantities of high-energy, extreme ultraviolet photons which illuminated the shell of gas that had been previously expelled when it shed its outer layers. This irradiated shell produced a glorious halo of excited gas that reemitted the extreme ultraviolet light from the hot stellar core in the form of the red light of hydrogen, the green light of ionized nitrogen, and the blue light of excited neutral oxygen. It had become, for a brief moment of cosmic time, a planetary nebula appearing as a ring of colorfully diffuse light that was being illuminated by a tiny hot blue star. An extraterrestrial "Herschel" scanning ancient skies would have easily noticed it as a nebula, not unlike the famous Ring Nebulae in the constellation of Lyra. In just a few hundred thousand years the nebula dimmed as it expanded and dispersed. The white dwarf continued cooling, rapidly at first, but slower and slower as its temperature steadily diminished. As it cooled the small traces of heavy elements present in the star's outer layers—such as iron, silicon, oxygen, nitrogen, and carbon—began to sink deeper into the atmosphere of the star, under the influence of the high gravitational field. After another 100 million years the star had cooled to 50,000 K, and the atmosphere was finally cleansed to the point that only one atom in a million heavier than hydrogen remained. The resulting pure hydrogen atmosphere was in reality only a thin veneer, consisting of perhaps one ten-thousandth or less of the mass of the white dwarf.

The star, like seven out of eight stars, which go through the planetary nebula phase, had become a pure hydrogen DA white dwarf star. The one in eight white dwarfs that contain no hydrogen in their outer layers evolve somewhat differently and end up having pure helium atmospheres. In an additional 40 million years the temperature of the star had dropped to 25,000 K to become the Sirius B first seen by Alvan Clark in 1862. It had shrunk to a diameter slightly smaller than that of the earth but with a mass equal to that of the sun (Figure 13.1, see also color section). The forces of gravity were now nearly five hundred thousand times that which we experience on the earth. Even though little or no energy was being produced in the interior of the star, it effectively maintained its high temperature for a very long period of time. The reason is that the only way it can lose heat is to emit radiation. However, because the star is so small and has such a relatively tiny surface area from which to radiate, it loses its thermal energy very slowly. The small surface of the star is effectively a bottleneck to energy loss, just as the water behind a huge reservoir would take centuries to empty from a small leak. Because radiation is virtually the only mode of energy loss and we can observe and calculate the amount being lost, this means that white dwarfs cool at rather predictable rates. They are effectively hourglasses, which we can use to measure the lengths of time since they were born. For example, using an updated version of a white dwarf cooling theory originally developed by the British astrophysicist Leon Mestel in 1952, it can be calculated that Sirius B has cooled by about eight-thousandths of a degree K and shrunk in radius by about 0.3 millimeter since it was first discovered over 144 years ago.

Today's Sirius B will continue to quietly cool. Eventually, in another billion years, its surface temperature will have dropped to about 13,100 K. At that point the star will

Figure 13.1. An artistic rendition of Sirius A and B (Space Telescope Science Institute).

begin to quiver. Deep in the stellar mantel, well above the degenerate core, a delicate balance will have been reached in which the ions of the hydrogen gas are at a critical temperature and pressure, where they begin to readily recombine. A slight cooling of the hydrogen in this zone tips the balance and results in a dramatic increase in the recombination of electrons and protons. The resulting hydrogen atoms readily absorb a fraction of the photons of ambient light, momentarily interrupting the steady outward flow of radiant energy from the stellar interior. This causes a build-up of energy in these layers, which in turn raises the temperature, reionizing the hydrogen atoms and releasing the pent-up energy. A temperature-sensitive valve will have been established. This hydrogen ionization valve modulates the flow of only a few percent of the star's energy, but that is enough to drive pulsations in the star with periods on the order of 2 to 20 minutes. On the surface, these pulsations manifest themselves in a complex array of small undulations, somewhat like those in a quivering water balloon. (These pulsations should not be confused with a "pulsar", which is a rapidly spinning neutron star.) The most noticeable effect is a rich pattern of constantly shifting temperature differences on the stellar surface resulting from these undulations. Viewed from a distance, so that it only appears as a point source, the temperature differences produce minute fluctuations in the stellar brightness of a few hundredths of a magnitude. The star will have reached what is called its ZZ Ceti phase, named after a particular white dwarf in the constellation Cetus, the Whale, in which such optical pulsations were first observed. (The ZZ designation comes from the practice of assigning variable stars in each constellation a one or two letter name, based on the order of their discovery.) A future astronomer who might happen to observe and record many hours of the pulsations from the future Sirius B would note that they were composed of many different discrete frequencies. A detailed analysis of the frequencies would reveal clues to the mass of Sirius B, its internal structure, and even its rotation rate in much the same way that geophysicists can determine the structure of the earth from the study of seismological waves produced by earthquakes. The pulsations will continue for several hundred million years, until the surface temperature has fallen to around 11,500 K, and the hydrogen ionization valve ceases to function effectively. At this point the pulsations die out and the star resumes its slow steady cooling.

All DA white dwarfs (those with pure hydrogen atmospheres) go through this pulsational phase as they cool through the range of temperatures between about 13,000 and 11,000 K. Such pulsations have actually been observed and studied in many such stars and the details of each star's set of pulsation frequencies are controlled primarily by its mass, its internal composition, and its rotation. By studying the pulsations, astronomers can actually deduce the mix of carbon and oxygen remaining in the degenerate core of the star. The pulsations can be remarkably stable. In the case of ZZ Ceti, 31 years of observations of the precise timing and counting of the pulsation cycles of the star's dominant 213-second pulsational mode has revealed that the pulsations are slowing down at a rate such that ZZ Ceti will lose one of its beats in about 1.2 billion years. This makes it one of the most stable known clocks in the universe. As ZZ Ceti cools, the amount of degenerate matter in the stellar interior grows, which in turn causes the pulsational period to lengthen. Measuring the slow frequency drift of the pulsation cycles is thus a measure of the rate at which ZZ Ceti is cooling, about 5 millionths of a degree Kelvin per year. This is one of the very few ways in which astronomers can actually "observe" the evolution of a particular star on the time scale of a human lifetime.

Sirius A will eventually retrace much of the life cycle of Sirius B, but at a much more leisurely pace. The primary difference between the evolution of Sirius A and the progenitor of Sirius B is that a degenerate helium core will begin forming well before Sirius A starts to run out of hydrogen fuel. When the time finally comes, perhaps in another 660 million years, the ignition of helium will be an abrupt, almost explosive event. It may begin in a localized region off center from the core in an episode called a "helium flash". It will not exit its hydrogen-burning phase for years. When it does, Sirius will once more appear as a luminous red star. After a brief period, however, another planetary nebula will form and dissipate and a new white dwarf will be born. Sirius A will lose only about two-thirds of its mass and the resulting white dwarf will be reduced to about 0.6 solar masses. A double degenerate system, one composed of two white dwarfs, will have formed, with Sirius B now the more massive and dominant component, being orbited by Sirius A in an expanded orbit. The two stars will orbit each other at a mean distance of 37 astronomical units, about the distance of Pluto from the sun, with an orbital period of about 180 years. From a distant vantage point

Figure 13.2. A double degenerate system containing two white dwarfs (indicated by the letters A and B). The two white dwarfs (LTT 12749 and LP 549-032) were discovered by Willem Luyten and are at a distance of 95 light years (image from the Sloan Digital Sky Survey, SKYVIEW).

like the earth, they would appear as two bluish stars one slightly brighter than the other traveling together through space; one of the countless millions of such systems of dead stars in our galaxy (Figure 13.2).

The two white dwarfs will continue to cool. The slightly larger, less massive, former Sirius A will cool faster than the smaller and more massive Sirius B, due to its lower heat capacity and larger surface area. Sometime, about 4 billion years in the future, the two companions, cooling at different rates, will have reached equal temperatures. This will occur when the two stars achieve surface temperatures of about 6000 K. Both will have lost their bluish white hue and will resemble the color of the sun. As cooling continues their luminosities will also fall steadily. The two stars will now be very faint and scarcely noticeable among the host of brighter stars which are still burning their hydrogen. At this time things will begin to change in a fundamental way in the degenerate core of the more rapidly cooling Sirius A. In spite of the very high temperature in the core of Sirius A, the carbon and oxygen nuclei will begin to "freeze". The bare nuclei will begin to assume a fixed lattice configuration, in effect becoming a solid. This transformation from a fluid state to a solid has been described by astronomers as the carbon atoms in the core forming a giant massive "diamond in the sky", in the sense that the core will become a crystalline form of carbon. The analogy with a real diamond, however, overlooks some rather "massive" differences. The matter in the core is still a degenerate hyper-dense mixture of carbon and oxygen nuclei. A 1 cubic centimeter "gem stone" of this putative diamond would weigh nearly 2.5 metric tons. In effect, a 5 billion year option on the mineral rights to the core of Sirius A is not a wise investment. A more apt analogy is that of a dense inaccessible form of matter resembling a very conductive metal. The matter in the core is still degenerate, with the free electrons continuing to hold back the gravitational pressure of the star. The change in the phase from the fluid state of free nuclei to a "solid", where the nuclei exist in a fixed lattice configuration, would be associated with the release of a relatively small amount of energy which would serve to slow, but not reverse, the rate of cooling of the white dwarf.

After 10 billion years the surface temperature of both stars will have fallen below 4000 K and they will become distinctly red. Hydrogen molecules can now form in the stellar atmospheres, absorbing some of the red light, which will eventually help give both stars a slightly more bluish tint. However, these cooler white dwarfs will scarcely be visible. They will have finally joined the countless other dim stellar corpses from past generations of the stars in our galaxy. The hunt for examples of such stars is currently a major quest of stellar astronomy. If such faint stellar corpses exist in great numbers then perhaps they contribute to the so-called dark matter that holds the outer portions of our galaxy together.

Possible evidence that such unseen stars exist comes from observations of what are called gravitational lenses. In the early 1990s astronomers finally achieved the ability to frequently monitor the brightness of millions of stars in a single night. The means to accomplish such Herculean measurements is due to the advent of large format CCD (Charged Coupled Devices) detectors, similar to those available in large-format electronic cameras having millions of pixels. Although the astronomical CCD detectors are much more light-sensitive and precise in their measurement of light, they operate in much the same way. In a single brief exposure of dense crowded star fields near the core of our galaxy or of our galaxy's close companions, the Magellanic Clouds, images of hundreds of thousands of stars are captured at once. These images are then compared with identical images obtained days or hours later. If a particular star appears to brighten between exposures, its behavior can be tracked over time. An abrupt brightening, followed by an identical decline in brightness, is a sign of a gravitational lens. The physical scenario is relatively simple. Imagine a massive dark star passes more or less directly across the line of sight of a random more distant star as seen from the earth. As the light rays from the more distant star pass near the dark intervening star, the unseen star's gravity bends the light rays, effectively focusing them. This bending is the same effect that Arthur Eddington measured during the 1919 solar eclipse which established Einstein's Theory of General Relativity. If the earth happens to pass near where these rays focus, the image of the distant star appears to brighten. If this brightening is followed by a symmetrical decline then the chances are good that a gravitational lens event has been observed. The alignments of the distant star, the dark body, and the earth must be precise, which effectively means that such events are rare, and frequently last a brief period of days or months. This is why large numbers of stars need to be observed to identify a handful of events. If enough such events are observed in a given direction then it is possible to statistically estimate the range of masses of the unseen bodies. The answer is: that in the direction of the central bulge of stars surrounding the center of our galaxy, the mean mass of these unseen bodies is about one-half a solar mass, just about that predicted for a large population of old, dim white dwarfs. The sum total of such objects, it now seems, cannot be sufficient to explain the large amount of dark matter that holds our galaxy and other galaxies together.

It is often stated that the sun, like Sirius A and B before it, will also one day become a red giant star and swell to such an enormous size that the earth will be engulfed, effectively ending the saga of life on earth. Although this destiny is seemingly inescapable, we are comforted by the fact that it remains a distant prospect, some 5 billion years in the future. That the sun will someday become a red giant is not in doubt, but the exact fate of the earth and the sequence of events leading up to this final stage are a matter of some debate. Will the earth be engulfed and destroyed, or will it survive as a charred cinder, forever orbiting the white dwarf that will form from the core of the deceased sun? The answer is that the outcome depends critically on many factors that are not well known. The fate of the interior planets Mercury and Venus, is sealed, they will be engulfed. The outer planets of Mars, Jupiter, and Saturn and beyond will survive. It is only the earth for which the final outcome is delicately poised.

The sun is presently some 4.58 billion years old, an age that can be quite accurately measured from the analysis of the radioactive elements present in certain types of meteorites. This age refers to the final phases of the formation of the solar system and marks the formation of the earth, as well, as the point at which the nuclear fires in the core of the sun were first lit. Since that time the sun has settled down as a hydrogen-burning main sequence star, steadily converting hydrogen into helium at an almost constant rate. On closer inspection, however, this burning is slowly producing changes in the core of the sun that are causing the sun to imperceptibly heat up and expand its outer radius. As hydrogen is consumed and the abundance of helium increases, the core begins to contract and the zone of hydrogen burning expands outwards. The net result is that the overall energy output increases slightly, which in turn modestly expands the solar radius. At present these changes are imperceptible since the solar radius is increasing at an approximate rate of about 2% per billion years. However, calculations show that in approximately 6 billion years this process will dramatically accelerate and in the relatively brief span of a billion years the sun will increase to ten times its present radius and enter the red giant phase of evolution. In the shrunken core, temperatures and pressures will rise and the nuclear burning zone will move further outward. The nuclear reactions and the central energy output of the sun will jump to levels several thousand times its present output. The outer layers of the sun will respond by dramatically expanding to over one hundred times its present radius. This expansion will cause the surface temperature to fall from approximately 5800 K to 3200 K. The sun will have become a red giant. If it were possible to view the sun from the present distance of the earth, the swollen sun would appear as a huge looming red disk filling nearly the entire sky. The solar surface would have a curious three-dimensional appearance, with huge convection cells of hot gas and magnetic storms slowly churning the visible surface. At the same time as this is occurring, the sun will begin rapidly shedding its outer layers in a fierce stellar wind, which peaks at a rate which will carry away as much as 1% of the sun's mass in ten thousand years. The sun will remain in this phase for perhaps 100 million years.

As a result of this final phase of its active evolution, the sun will have shed some 45% of its mass. It will also develop a short-lived planetary nebula that will announce to the galaxy the demise of yet another G star. The resulting white dwarf remnant will have a mass of 0.55 solar masses and will spend several more billion years cooling to temperatures at which it will no longer be visible. The final death throws of the sun will, however, have a significant impact on its retinue of planets, the earth in particular.

The earth is not simply a passive recipient of the sun's radiation, it will respond with changes that will dramatically affect life on this planet. Initially the rising temperature of the sun will be compensated by biological and geochemical processes, which seek to moderate rising temperatures on earth. Ultimately, however, surface temperatures on the earth will inevitably rise and life on this planet will undergo dramatic changes. These processes begin with the evaporation and ultimate loss of the earth's oceans. Presently the earth is losing its water through a slow photochemical leak into interplanetary space. This happens imperceptibly in a process that begins with the evaporation of water at the earth's surface. As water vapor rises most is trapped in a thermal inversion layer at the base of the troposphere, about 12 km above the earth's surface. A small amount, however, enters the stratosphere where solar ultraviolet radiation dissociates or breaks down the water molecules into hydrogen and oxygen. The lighter hydrogen atoms naturally rise in the atmosphere forming an exosphere that extends outwards several earth radii. This extensive hydrogen exosphere, which glows due to scattered light from solar Lyman alpha photons, was first photographed, in ultraviolet light by Apollo 16 astronauts from the surface of the moon in 1972, as a large corona extending out almost two earth diameters on the day side. This emission is best known to astronomers as the faint fog of atomic hydrogen ultraviolet emission lines that contaminate long exposures from earth-orbiting satellite observatories such as *Hubble* or *FUSE* that view the universe from deep within the earth's enveloping shroud of escaping hydrogen. The corona, although large, contains relatively little hydrogen and most of it is permanently bound to the earth. Only a small fraction escapes into interplanetary space, where it is forever lost. This escape of hydrogen represents a net loss of water from the earth's oceans. At present the loss is insignificant, only a millimeter per million years decrease in the sea level.

As the surface temperature of the earth increases, however, all the factors that drive this water loss process will increase and the loss of the oceans will accelerate dramatically. A critical point will be reached when the earth's surface temperature rises to approximately 60°C (140°F) and life on earth as we know it will become untenable. Animal and plant life will disappear and the earth will once again revert to a microbial flora, similar to that from which it originally sprang. The effect of higher temperatures on the earth's atmosphere will be equally dramatic. The water vapor content of the earth's stratosphere will increase and even more hydrogen will be lost to space. The increasing water vapor will also lead to what is called the "moist greenhouse" in which the higher water vapor content in the atmosphere will trap solar radiation increasing the earth's surface temperatures, leading to ever more rapid evaporation rates. An irreversible positive feedback will drive the surface temperature higher and produce even faster evaporation and more rapid hydrogen loss. This phase is relatively short-lived, since the oceans could evaporate in as little as one hundred million years. In the final stages, what little surface water remains would be concentrated into shrinking alkali lakes like those of the East African Rift Valley and finally into hot briny ponds and streams like those seen in some volcanic calderas. The only life possible at the surface would be thermopilic and halophilic bacteria. The exposed surface of the earth would be a barren desert with lofty continental highlands and extensive deep abysses which once contained the oceans. The last vestiges of life on earth will retreat into the rocks themselves. Eventually, however, as the sun continues to grow and heat the earth all life will finally be extinguished at around 300°C (572°F) and certainly by 374°C (705°F), when water in liquid form can no longer exist. As the temperature increases further the carbon dioxide locked in the earth's limestone will be released and a runaway CO_2 greenhouse, like that on Venus, will exist. Eventually surface temperatures on the earth will reach 1000°C (1800°F) and its surface rocks will begin to melt, creating a vast featureless ocean of lava.

The sun is not yet finished, but things now begin to happen very swiftly on an astronomical time scale. As hydrogen is exhausted in the solar core, it shrinks further and temperatures and pressures surge; helium begins to fuse into carbon and oxygen, further increasing the energy output of the sun. Now it is on a very fast track. The sun's outer layers will swell dramatically absorbing first Mercury then Venus until its thin wispy atmosphere laps at the earth's orbit. The earth may well survive this enlargement, since the red giant phase is also associated with the loss of mass, and the earth will spiral outward in its orbit due to the weakening gravitational pull of the sun. Perhaps through this orbital retreat from the sun the earth may just elude being engulfed. The helium-burning phase is also unstable, and the sun will undergo a series of ever more catastrophic enlargements and contractions, called thermal pulses, that will last for a period of just over 500,000 years. During one of these thermal pulses the chance exists that the earth will be engulfed and spiral into the sun where our planet will make a small contribution to the heavy elements in the sun's atmosphere.

The earth could also escape this fate and continue to orbit the white dwarf that will finally become the sun. If it does, it will cool and become a lifeless slag heap endlessly orbiting a faint, slowly cooling star that would appear as a tiny, but brilliant point of light in an airless sky. All traces of the earth's nearly 12 billion year history will have been obliterated.

Appendix A

Glossary

Aberration of Starlight describes a small apparent displacement of the position of a star due to the observer's motion perpendicular to the line of sight. Discovered by James Bradley in 1728, stellar aberration is the effect of the earth's orbital velocity (30 km/sec) causing the position of a star to shift by up to 40 arc seconds opposite to the direction of the earth's motion. This shift is due to the finite speed of light and was the first physical demonstration of the earth's orbital motion.

Arc Second One arc second is an angle corresponding to 1/3600th (1/60th of 1/60th) of a degree. It is the angular unit in which stellar parallaxes are measured. The angular diameter of the earth's orbit at a distance of 1 *Parsec* is 1 arc second.

Atomic Mass Unit (amu) is the basis of the system of atomic masses. A carbon atom has a mass of 12 amu. The hydrogen atom has a mass of 1.008 amu. One amu $= 1.66 \times 10^{-24}$ grams.

Atmospheric Refraction is the bending of the path of a ray of the light from a celestial body by the earth's atmosphere. As light enters the earth's atmosphere, at an angle to the vertical, the increasing atmospheric density causes a slight downwards bending of the light path towards the center of the earth. This in turn causes the apparent position of a body to appear at a higher elevation. The magnitude of the effect increases from essentially no shift at the zenith to larger shifts near the horizon.

Balmer Lines The series of absorption or emission lines in the visible part of the spectrum produced by atomic hydrogen, due to electron transitions which originate or terminate with the second lowest energy level. The first three longest wavelength members of the Balmer series—labeled $H\alpha$, $H\beta$, and $H\gamma$—are located at wavelengths of 6568 Å, 4861 Å, and 4340 Å. Johann Jacob Balmer in 1885 established a simple empirical relation between the Balmer lines based on integers.

Brown Dwarf A low-luminosity star-like body whose mass is too low (less than approximately 0.08 solar masses) to initiate hydrogen burning in the core. Brown dwarfs, which emit residual thermal energy derived from gravitational contraction, have surface temperatures of 2000 K, or lower, and emit most of their light at infrared wavelengths. Brown dwarf spectral types include L and T dwarfs.

Lyman Lines The series of absorption or emission lines in the far ultraviolet part of the spectrum produced by atomic hydrogen, due to electron transitions which originate or terminate with the lowest energy level. The first three longest wavelength members of the Lyman series—labeled $L\alpha$, $L\beta$, and $L\gamma$—are located at wavelengths of 1216 Å, 1026 Å, and 972 Å. Theodore Lyman in 1906 discovered the Lyman series by observing electron-bombarded hydrogen in a vacuum.

Celestial Coordinates (also called equatorial coordinates) are a latitude–longitude system based on the plane of the earth's equator and its rotational pole. *Declination* is the angle measured (in degrees) north and south of the celestial equator and *Right Ascension* is the angle, measured eastward from the first point of Aries, along the celestial equator. *Right Ascension* is often measured in units of time (hours, minutes, and seconds), since the sky appears to make one complete cycle every 24 hours. Because of **Precession** the *Right Ascension* and *Declination* of a star will not remain fixed, thus an epoch such as the year 2000.0 needs to be specified.

Declination (*see* **Celestial Coordinates**)

Ecliptic The apparent path which the sun follows with respect to the stars during the year. Consequently it is also the plane of the earth's orbit projected onto the celestial sphere. The plane of the earth's equator is tilted by 23.5 degrees with respect to the ecliptic.

Ecliptic Coordinates are a latitude–longitude system in which ecliptic latitudes are measured as positive degrees north of the ecliptic and negative degrees south of the ecliptic. Degrees of ecliptic longitude are measured eastward from the vernal equinox or the (ascending) point of intersection of the plane of earth's equator with the plane of the ecliptic, traditionally called the first point of Aries. Due to the **Precession of the Equinoxes**, the origin of the first point of Aries moves westward at a rate of 50.3 arc seconds per century. Two millennia ago the first point of Aries was located in the constellation Aries, it is presently in Pisces.

Equation of State For a gas or fluid the equation of state is the mathematical relation between the pressure and other properties such as the density and temperature. For example, the ideal gas law states that the pressure is proportional to the density and temperature of a gas.

General Theory of Relativity The theory published by Albert Einstein in 1916 that treats gravitation and acceleration. It postulates that uniform acceleration and gravitational forces are equivalent phenomena. This leads to a description of the gravitational force of a body as arising from a deformation of space and time produced by the body's mass. Among the consequences of General Relativity, at the stellar level, are the bending of the path of starlight near a massive body, the gravitational redshift, and black holes.

Heisenberg's Uncertainty Principle A fundamental quantum mechanical property which establishes that there is a limit to the joint knowledge of the momentum and the position of a particle. The more precisely the momentum is defined the less precisely the position will be known and vice versa.

Heliometer A specialized refracting telescope designed to precisely measure the solar diameter and angles between stars. The objective lens of a heliometer is cut in half and mounted so that the two halves can be slid laterally with respect to each other. Each half produces a separate image of a stellar field and these two images can be moved by a micrometer screw driving the lens halves. Friedrich Bessel used a heliometer to measure the first parallax of a star in 1838.

Kepler's Laws of Planetary Motion Three empirical laws derived by Johannes Kepler, between 1609 and 1618, that accurately describe the motions of the planets about the sun. Kepler derived the first two laws from an intensive theoretical study of the positions of the planet Mars observed by Tycho Brahe. The first law states that the planets move in elliptical orbits with the sun at one center, or foci. The second law states the radius vector from the sun to the planet sweeps out equal areas in equal times. This law specifies how the orbital speed of a planet increases as its heliocentric distance decreases and vice versa. The third law relates the distance of a planet and its orbital period: the cube of the distance is proportional to the square of its orbital period. In modern terms, the first and third laws are consequences of Newton's inverse square law of gravitation. The second law is a consequence of the conservation of angular momentum.

Newton's Laws of Motion Three statements of the relationship between mass, force, acceleration, and inertia on which Isaac Newton based his 1686 *Principia.*

1. *A body at rest will remain at rest and a body in uniform motion will remain in that state, unless acted on by a force.*

The second law provides a mathematical definition of the concept of "force", and relates this force to the mass of a body through the acceleration it produces.

2. *The acceleration in the motion of a body is proportional to the force acting on it and will occur in the direction of the force.*

This is most often summarized in the equation $F = ma$, where F is the force, m is the mass, and a is the acceleration.

The third law is required to complete the concepts embodied in the first two laws.

3. For every action there is an equal and opposite reaction.

This means that if you push on a body, it pushes back.

Parallax is the apparent angular shift in the position of a distant object when it is viewed from two different positions perpendicular to the line of sight. For the stars, the relative parallax shift is a small apparent angular shift of a foreground star with respect to much more distant background stars, when viewed from opposite sides of the earth's orbit. The corresponding absolute parallax is the shift with respect to a point of reference effectively at an infinite distance. The nearest stars have parallaxes of less than 0.75 seconds of arc. The unit of stellar distances, the *Parsec* (3.26 light years), is defined such that a hypothetical star at a distance of 1 parsec would have a parallax of 1 arc second. The relationship between the parallax in arc seconds, π, and the distance d in parsecs is $\pi = 1/d$. There is also a much smaller parallax, the diurnal parallax, due to the angular shift of a nearby object, such as the moon, when viewed from different points on the earth's surface.

Parsec (*see* **Arc Second**)

Polytrope In astronomy a polytrope refers to a solution to the Lane–Emden equation (J. Homer Lane [1819–1880] and Robert Emden [1862–1940]), which describes the equilibrium properties of a self-gravitating sphere of gas. A polytrope can be thought of as a highly idealized model of a stellar interior which meets the criteria of hydrostatic equilibrium and has an equation of state which is a power law of its density.

Precession is a motion of the earth, in which the earth's rotational pole slowly circulates in a conical path with respect to the stars. This motion is analogous to the conical motion of a spinning top, whose spin axis is not aligned with the vertical. In the case of the earth, the radius of the conical motion is 23.5 degrees. It requires approximately 25,800 years to complete a full precessional circuit. Precession is caused by the combined gravitational torque of the sun and the moon on the earth's equatorial bulge.

Precession of the Equinoxes was discovered by Hipparchus in the 2nd century BC and consists of the slow eastward drift of the equinoxes with respect to the constellations. The motion of the earth's equator with respect to the plane of the earth's orbit (the ecliptic) results from **Precession**. The precessional rate is 50.3 arc seconds per century, which equates to the equinoxes making a complete circuit of the constellations of the zodiac in 25,800 years.

Ptolemaic System The geocentric theory of planetary motions devised by the 2nd century AD Greco-Egyptian astronomer Claudius Ptolemy. The center of Ptolemy's system was a stationary earth. All observed celestial motions were mathematically described by a set of uniform circular motions about the earth. For the outer planets (Mars through Saturn) each planet moved uniformly along a large circular path called a deferent, whose center was offset from the earth by a distance called the equant. As the planet moved along the deferent it simultaneously moved at a uniform rate around a smaller circle called an epicycle, which was centered on the circumference of the deferent. This system of compound motions was capable of approximating the observed motions of the planets. The fixed stars occupied an exterior eighth sphere.

Right Ascension (*see* **Celestial Coordinates**)

Special Theory of Relativity The theory published by Albert Einstein in 1905 that stated that the speed of light in a vacuum would always be measured as a constant by any observer, regardless of the state of motion. This theory dealt primarily with uniform motions. Among the consequences are that time intervals, distances, and masses are measured differently, depending on the observer's state of motion. Special Relativity is the source of Einstein's famous equation $E = mc^2$.

Appendix B

Properties of the Sirius System

SIRIUS

Names:	The Dog Star, Alpha Canis Majoris, 9 CMa, HR 2491, HD 48915, BD –16° 1591, ADS 5423, AC 1, LHS 219
Position:	$\alpha_{2000} = 06^h\ 45^m\ 08.9173^s$
	$\delta_{2000} = -16°\ 42'\ 58.017''$ (epoch 2000)
Proper Motion:	$\mu_\alpha = 0.5461''/\text{yr}$
	$\mu_\delta = -1.2231''/\text{yr}$
	$\rho = 1.3395''/\text{yr}$
	$\theta = 204.0576°$
Distance:	2.6314 pc, 8.553 light years, 8.120×10^{13} km
Parallax (π):	380.023 ± 1.283 milli-arc seconds
Radial Velocity:	-8.7 km/s

ORBIT

Period:	50.075 ± 0.103 years
Epoch:	1994.352 ± 037 (last periapse passage)
a:	$7.50 \pm 0.03''$
	19.82 AU
e:	0.59657 ± 0.00130 (Eccentricity)
i:	$138.4376° \pm 0.298°$ (Inclination)
Ω:	$150.334° \pm 0.348°$ (Longitude of Ascending Node)
ω:	$45.685° \pm 0.240°$ (Argument of Periapse)

SIRIUS A

V Magnitude:	−1.46
Spectral Type:	A1V
Mass:	$2.02 \pm 0.03\ M_{\odot}$
Radius:	$1.712 \pm 0.00025\ R_{\odot}$
Mean Density:	0.568 grams/cm^3
Luminosity:	$25.4 \pm 1.3\ L_{\odot}$
Temperature:	10,500 K
Gravity:	8.556 ± 0.010 (log of Surface Gravity)
Age:	238 ± 13 million years (Main Sequence)

SIRIUS B

WD0642-162

V Magnitude:	+8.44
Spectral Type:	DA2
Mass:	$1.00 \pm 0.01\ M_{\odot}$
Radius:	$0.0084 \pm 0.00025\ R_{\odot}$ (5840 km)
Mean Density:	2.38×106 grams/cm^3
Central Density:	32.36×106 grams/cm^3
Temperature:	$25{,}193 \pm 37$ K
Gravity:	8.556 ± 0.010 (log of Surface Gravity in cgs units)
Vgr:	$+80.42 \pm 4.83$ km/s (Gravitational Redshift)
Age:	123 million years (White Dwarf Cooling Age)
Progenitor Mass:	$5.056\ M_{\odot}$ ($\pm 7\%$)

References

ABBREVIATIONS

A&A	Astronomy and Astrophysics
AJ	Astronomical Journal
AN	Astronomische Nachrichten
ApJ	Astrophysical Journal
ApJL	Astrophysical Journal Letters
ApJS	Astrophysical Journal Supplement
JBAA	Journal of the British Astronomical Association
JHA	Journal of the History of Astronomy
MNRAS	Monthly Notices of the Royal Astronomical Society
Obs.	The Observatory
PASP	Publications of the Astronomical Society of the Pacific
Phil. Trans.	Philosophical Transactions of the Royal Society
Q. Jl. Astr. Soc.	Quarterly Journal of the Royal Astronomical Society

CHAPTER 1

Ceragioli, Roger C., 1993, *The Riddle of Red Sirius: An Anthropological Perspective, in Astronomies and Cultures*, eds. Clive L. N. Ruggles and Nicholas J. Saunders, University of Colorado Press, 67–99.

Evans, James, 1998, *The History & Practice of Ancient Astronomy*, Oxford University Press, 166–177.

Ingham, M. F., 1969, *The length of the Sothic cycle*, Journal of Egyptian Archaeology, lv, 36–40.

Krupp, Edwin C., 1977, *In Search of Ancient Autonomies*, Doubleday & Co.

Lockyer, J. Norman, 1894, *The Dawn of Astronomy, Reprint edition*, Cambridge, MA, MIT Press, 1973.
Members of the David H. Koch Pyramids Radiocarbon Project, 1999, *Dating the Pyramids*, Archeology, 52, 5, 27–33.
Michael R. Molnar, 1995, *Sirius Rising: Commemorating the Anniversary of the Sothic Cycle*, Journal of the Society for Ancient Numismatics, Vol. XIX, No. 1, 15–18.
Redford, Donald B., 2001, *Oxford Encyclopedia of Ancient Egypt*.
Schaefer, Bradley E., 2000, *The Heliacal Rise of Sirius and Ancient Egyptian Chronology*, JHA, 31, 149–155.
See, T. J. J., 1892, *Explanation of the Mystery of the Egyptian Phoenix*, Astronomy and Astro-Physics, 11, 457.
van Gent, Robert, H., 1989, *The Colour of Sirius*, Obs., 109, 23–24.
Wells, Ronald A., 1993, *Origin of the Hour and Gates of the Duat*, Studien zur Altägyptischen Kultur, 20, 305–326.
Wells, Ronald A., 1985, *Sothis and the Satet Temple on Elephantine: A Direct Connection*, Studien zur Altägyptischen Kultur, 12, 255–302.

CHAPTER 2

Allen, Richard Hinckley, 1963, *Star Names Their Lore and Meaning*, Dover Publications, 117–129.
Aratus, *Phaenomena*, 1997, translated by Douglas Kidd, Cambridge University Press, 97–98.
Aveni, Anthony, F., 1981, *Tropical Astronomy*, Science, 213, 161–171.
Burnham, Robert Jr., 1963, *Burnham's Celestial Handbook*, Dover Publications, 386–415.
Ceragioli, Roger C., 1995, *The Debate Concerning 'Red Sirius'*, JHA, 26, 187–227.
Ceragioli, Roger C., 1996, *Solving the Puzzle of 'Red Sirius'*, JHA, 27, 93–128.
Ceragioli, Roger C., 1993, *The Riddle of Red Sirius: An Anthropological Perspective, in Astronomies and Cultures*, eds. C. L. N. Ruggles and N. J. Saunders, University of Colorado, 67–99.
Chamberlain, Von Del, 1982, *When the Stars Came down to Earth: Cosmology of the Skidi Pawnee Indians of North America*, Ballena Press, Los Altos, CA.
Darmesteter, James, 1898, *Sacred Books of the East*, American Edition, Yasht 8, *Hymn to Sirius*.
Eddy, John A., 1974, *Astronomical Alignment of the Big Horn Medicine Wheel*, Science, 184, 1035–1043.
Evelyn-White, Hugh G., 1914, *Hesiod, The Homeric Hymns, and Homerica*, Loeb Classics, Cambridge, MA, 47.
Griaule, Marcel and Dieterlen, Germaine, 1965, *Le Renard pâle*, Tome I, Fasicule, 1; Institut d'Ethnologie, Musée de l'Homme, Palais de Chaillot, Place du Trocadero, Paris, 16e.
Homer, *The Iliad*, translated by Robert Fitzgerald, 1974, Anchor Press/Doubleday.
Jiang xiao-yuan, 1993, *The colour of Sirius as recorded in ancient Chinese texts*, Chinese Astronomy and Astrophysics, 17/2, 223–228.
Lewis, David, 1978, *The Voyaging Stars: An Account of Polynesia Astronomy*, Yale University Press, New Haven, CT.
Makemson, Maud Worchester, 1941, *The Morning Star Rises: An Account of Polynesia Astronomy*, Yale University Press, New Haven, CT.

Miller, Dorcus S., 1997, *Stars of the First People; Native American Star myths and Constellations*, Pruett.
Olcott, William Tyler, 1911, *Star Lore of All Ages*, G. P. Putnam's Sons, New York.
Reyahi, Bassel A., 1998, *Sirius: A Scientific and Qur'anic Perspective*, private communication.
Santillana, Giorgio de and Hertha von Dechend, 1969, *Hamlet's Mill: An Essay on Myth and the Frame of Time*, Gambit Inc., Boston.
Schaefer, Bradley E., 2004, *The Latitude and Epoch of the Origin of the Astronomical Lore of Eudoxus*, JHA, 35, 116–223.
Schaefer, Bradley E., 2005, *The Epoch of the Constellations on the Farnese Atlas and Their Origin in Hipparchus's Lost Catalogue*, JHA, 36, 167–196.

CHAPTER 3

Cook, Alan, 1998, *Edmund Halley charting the Heavens and the Seas*, Clarendon Press, Oxford.
Grant, Edward, 1997, *The Medieval Cosmos: Its Structure and Operation*, JHA, 28, 147–168.
Halley, Edmund, 1718, *Considerations on the Change of the Latitudes of Some of the principal fixt Stars*, Phil. Trans, 30, 736–738.
Halley, Edmund, 1720, *Of the Infinity of the Sphere of the Fix'd Stars*, Phil. Trans., 31, 22–24.
Halley, Edmund, 1720, *Of the Number, Order and Light of the Fix'd Stars*, Phil. Trans., 31, 24– 26.
Herschel, William Friedrich, 1782, *On the Parallax of the Fixed Stars*, Phil. Trans., 72, 492–501.
Herschel, William Friedrich, 1803, *Account of the Changes that have happened during the last Twenty-Five Years, in the relative Situation of Double-Stars with an Investigation of the Cause to which they are owing*, Phil. Trans., 93, 339–382 and 94, 353–384.
Hirshfeld, Alan M., 2001, *The Race to Measure the Cosmos*, W. M. Freeman & Co., New York.
Hoskin, Michael, 1985, *Stukeley's Cosmology and Newtonian Origins of Obler's Paradox*, JHA, 16, 77–112.
Kepler, Johannes, 1604, *Optics: Paralipomena to Wileo & Optical Part of Astronomy*, 2000, translated by William H. Donahue, Green Lion Press.
Newton, Isaac, 1728, *A Treatise of the System of the World*, translated by Bernard I. Cohen, 1969, Dawsons of Pall Mall, London.
White, Andrew Dickson, 1898, *A History of the Warfare of Science with Theology in Christendom*, D. Appleton & Company, New York.

CHAPTER 4

Auwers, Arthur, 1862, *On the Irregularities of the Proper Motion of Sirius*, MNRAS, 22, 147–50.
Auwers, Arthur, 1864, *Ueber die Bahn Des Sirius*, AN, 63, 273–84.
Auwers, Arthur, 1865, *Orbit of Sirius*, MNRAS, 25, 38, 40.
Baum, Richard and Sheehan, William, 1997, *In Search of the Planet Vulcan*, Plenum Press.
Bessel, Wilhelm Friedrich, 1844, *On the Variation of the Proper Motions of Procyon and Sirius*, MNRAS, VI, 136–144.
Bessel, Wilhelm Friedrich, 1839, *A letter from Professor Bessel to Sir. J. Herschel*, MNRAS, VI, 152–161.
Bessel, Wilhelm Friedrich, 1838, *Fernere Nachricht von der Bestimmung der Entfernung des 61 Sterns des Schwans*, AN, 16, 66–95.

Hirshfeld, Alan M., 2001, *The Race to Measure the Cosmos*, W. M. Freeman & Co. New York.
Holberg, J. and Wesemael, F., 2007, *The Discovery of the Companion of Sirius and Its Aftermath*, JHA (in press).
Fricke, Walter, 1985, *Freidrich Wilhelm Bessel (1784–1846)*, Astrophysics and Space Science, 110, 11–19.
Jones, Bessie Z., 1967, 1968, *Dairy of the Two Bonds: 1846–1849*, Harvard Library Bulletin, XV, 4; XVI, 1, 2, 385.
Peters, Christian A. F., 1851, *Ueber die eigene Bewegnung des Sirius*, Habilitation, Königsberg.
Safford, Truman Henry, 1862, *On the Proper Motions of Sirius in Declination*, MNRAS, 22, 145, 144–147.

CHAPTER 5

Alan Graham Clark, 1897, The Cambridge Chronicle, 12 June 1897.
Alvan Clark, Harper's Weekly, Sept. 3, 1887, 631.
Alvan Clark and his method of Object-Glass Working, 1887, The English Mechanic, No. 1174, 83.
Brashear, J. A., 1892, *George Bassett Clark*, Astronomy and Astro-Physics., 11, 367–372.
Bond, George P., 1862, *Letter to editor*, AN, 58, 85–90.
Burnham, Sherwood W., 1879, *Double Stars discovered by Mr. Alvan G. Clark*, American Journal of Science, 17, 283–289.
Church, John A., 1963, *Optical Designs of Some Famous Refractors*, Sky and Telescope, 63, 302–308.
Clark, Alvan, 1889, *Autobiography of Alvan Clark*, Sidereal Messenger, 8, 109–117.
Clark, Alvan, 1867, *In Receipt of the Rumford Medal*, Proceedings of the American Academy of Arts and Sciences, 7, 244–249.
Clark, Alvan, 1841, *On Rifle Shooting*, American Repertory of Arts, Sciences and Manufactures, III, 164–169.
Clark, Alvan G., 1893, *Great Telescopes of the Future*, Astronomy and Astro-physics 12, 673– 678.
Clark, Alvan G., 1893, *Possibilities of the Telescope*, North American Review, 156, 48–53.
Clarks Telescope Works, 1892, The Cambridge Chronicle, 2 January 1892.
Death of Alvan G. Clark, 1895, Popular Astronomy, 5, 167.
Fox, Phillip, 1915, *A General Account of the Dearborn Observatory*, Annals of the Dearborn Observatory, 1, 1–7.
Fulton, John, 1896, *Memoirs of Frederick A. P. Barnard*, Macmillan & Co., New York, 244–246.
Galignani's Messenger, 1862, Paris, April 1–2.
George Bassett Clark, 1891, Proceedings of the American Academy of Arts and Sciences, 27, 360–363.
Goldschmidt, M., 1863, *Le cortège de Sirius*, Les Mondes, 5.
Goldschmidt, Revd. M., 1863, *Companions of Sirius*, MNRAS, 23, 182.
Hawkins, William B., 1927, *The Clarks*, Popular Astronomy, 35, 379–382.
Holberg, J. and Wesemael, F., 2007, *The Discovery of the Companion of Sirius and Its Aftermath*, JHA (in press).
Knott, George, 1866, *On the Companion to Sirius*, MNRAS, 26, 243.
Lassel, W., 1862, *Letter to editor*, AN, 57, 251–252.

Leavitt, Henrietta S., 1896, *Clarks Observatory* in Cambridge Sketches, ed. E. M. H. Merril, Boston, 149–154.

Moigno, Abbé, 1862, Cosmos, March 28, p. 377, 391–393.

Newcomb, Simon, 1866, *Measures of the Companion of Sirius made at the U.S. Naval Observatory Washington in 1866, with a note on its identity with the disturbing body indicated by theory*, AN, 66, 381.

Newcomb, Simon, 1874, *New Refracting Telescope of the National Observatory, Washington D.C.*, Science Record, 324–332.

Newcomb, Simon, 1910, *Popular Astronomy*, Macmillan & Co. Ltd., London, 139–144.

Newcomb, Simon, 1873–1874, *The Story of a Telescope*, Scribner's Monthly, 7, 44–54.

Palmer, Charles S., 1927, *Two Hours with Alvan Clark Sr.*, Popular Astronomy, 35, 143–145.

Report of the Committee of the Overseers of Harvard College Appointed to Visit the Observatory, 1862.

Report of the Committee of the Overseers of Harvard College Appointed to Visit the Observatory, 1863.

Rutherfurd, Lewis, 1862, *Companion to Sirius*, American Journal of Science, 34, 294–295.

Schaeberle, J. M., 1896, *Discovery of the Companion to Procyon*, PASP, 8, 314.

Struve, M. Otto, 1864, *On the Satellite of Sirius*, MNRAS, 24, 149–152.

Struve, M. Otto, 1866, *On the Satellite of Sirius*, MNRAS, 26, 267–271.

Sullivan, John F., 1927, *A Visit to Alvan Clark, Jr.*, Popular Astronomy, 35, 388–391.

The Alvan Clark Establishment, 1887, Scientific American, 57, 198–199.

Wendell, Oliver C., 1897, *Alvan Graham Clark*, Proceedings of the American Academy of Arts and Sciences, 33, 520–524.

Warner, Debra J. and Ariail, R. B., 1995, *Alvan Clark & Sons, Artists in Optics*, Willman-Bell.

Williams, Thomas R., 1996, *The development of astronomy in the Southern United States*, JHA, 27, 13–44.

CHAPTER 6

Aitken, Robert G., 1896, *Measures of Sirius*, PASP, 8, 314, and AJ, 17, 265 (1896).

Aitken, Robert G., 1942, Sky and Telescope, September 3, Vol. 1, No. 11, 5.

Brown, Stimson J., 1896, *Observations of the Companion of Sirius*, AJ, 17, 46.

Clerke, Agnes M.; Fowler, A.; & Gore, John E., 1898, *Astronomy*, 487.

Campbell, William Wallace, 1905, *The Variable Radial Velocity of Sirius*, ApJ, 21, 176–184.

FitzGerald, A. P., 1966, *John Ellard Gore (1845–1910)*, The Irish Astronomical Journal, 7, 213– 218.

Hearnshaw, J. B., 1986, *The Analysis of Starlight*, Cambridge University Press.

Holden, Edward S., 1896, *Companion of Sirius*, AJ, 17, 26.

Gore, John Ellard, 1891, *The Companion of Sirius*, JBAA, 1, 318–319.

Knott, George, 1866, *On the Companion to Sirius*, MNRAS, 26, 243.

Schaeberle, J. M., 1896, *Discovery of the Companion to Procyon*, PASP, 8, 314.

See, T. J. J., 1896, *Rediscovery and Measurement of the Companion of Sirius*, AJ, 17, 1–2.

Struve, M. O., 1866, *On the Satellite of Sirius*, MNRAS, 26, 267.

CHAPTER 7

Adams, Walter S., 1914, *An A-Type Star of Very Low Luminosity*, PASP, 26, 198.

Adams, Walter S., 1915, *The Spectrum of the Companion of Sirius*, PASP, 27, 236–237.

Adams, Walter S. and Kohlschütter, Arnold, 1914, *Some Spectral Criteria for the Determination of Absolute Stellar Magnitudes*, ApJ, 40, 385–399.

Anderson, J. A., 1920, *Application of Michelson's Interferometer Method to the Measurement of Close Double Stars*, ApJ, 51, 263–275.

Barstow, Martin A., Bond, H. E., Burliegh, M. R. and Holberg, J. B., 2003, *Resolving Sirius-like binaries with the Hubble Space Telescope*, MNRAS, 322, 891–900.

DeVorkin, D. H., 2000, *Henry Norris Russell, Dean of American Astronomers*, Princeton University Press.

Eddington, Arthur S., 1920, *The Internal Constitution of the Stars*, Nature, 106, 14–20.

Eddington, Arthur S., 1924, *On the relation between the masses and luminosities of the stars*, MNRAS, 84, 308–333.

Eddington, Arthur S., 1926, *The Internal Constitution of the Stars*, Cambridge University Press.

Eddington, Arthur S., 1926, *Stars and Atoms*, Cambridge University Press.

Hernshaw, J. B., 1986, *The Analysis of Starlight*, Cambridge University Press.

Hertzsprung, Ejnar, 1905, *Zur Strahlungen der Sterne*, translation in Source Book in Astronomy 1900–1950, ed. Harlow Shapley, Harvard University Press, 1960, 248–252.

Hertzsprung, Ejnar, 1909, *On New Members of the System of the Stars* β, γ, δ, ε, ζ, *Ursae Majoris*, ApJ, 30, 135–145.

Hertzsprung, Ejnar, 1915, *Effective wavelengths of absolutely faint stars*, ApJ, 42, 111–119.

Hertzsprung to Adams, Correspondence November 14, 1914.

Lindblad, B., 1922, *Spectrophotometric methods for determining stellar luminosity*, ApJ, 55, 85– 118.

Lockyer, Norman J., 1883, *Elements of Astronomy*, D. Appleton & Co.

Luyten, Willem J., 1922, *Note on Some Faint Early-Type Stars with Large Proper Motions*, PASP, 34, 54–55.

Luyten, Willem J., 1922, *Additional Note on Faint Early-Type Stars with Large Proper Motions*, PASP, 34, 132.

Luyten, Willem J., 1922, *Third Note on Faint Early-Type Stars with Large Proper Motions*, PASP, 34, 356–357.

Luyten, Willem J., 1923, *A Study of Stars with Large Proper Motion*, Lick Observatory Bulletin, Vol. 11, No. 344, 1–32.

Luyten, Willem J., 1987, *My First 72 Years of Astronomical Research: Reminiscences of an Astronomical Curmudgeon*, privately published.

Michelson, Albert A. and Pease, F. G., 1921, *Measurement of the Diameter of Orionis with the Interferometer*, ApJ, 53, 249–259.

Russell, Henry N., 1913, *"Giant" and "Dwarf" Stars*, Obs., 36, 324–329.

Russell, Henry N., 1914, *Relations between the Spectra and other Characteristics of the Stars*, Popular Astronomy, 22, 275–351.

Russell, Henry N., 1944, *Notes on White Dwarfs and Small Companions*, AJ, 22, 13–17.

Upgren, A. R., 1995, *Willem Jacob Luyten (1899–1994)*, PASP, 107, 603–605.

van Maanen, A., 1917, *Two Faint Stars with Large Proper Motions*, PASP, 29, 258–259.

Vibert, Douglas A., 1957, *The Life of Arthur Stanley Eddington*, Thomas Nelson & Sons Ltd.

Young, Charles A., 1891, *General Astronomy*, Ginn & Co.

CHAPTER 8

Chandrasekhar, S., 1983, *Eddington: The Most Distinguished Astrophysicist of This Time*, Cambridge University Press.

Chandrasekhar, S., 1939, *An Introduction to the Study of Stellar Structure*, Dover.

Chandrasekhar, S., 1934, *Stellar Configurations with Degenerate Cores*, Obs., 57, 373–377.

Chandrasekhar, S., 1931, *The Density of White Dwarf Stars*, Philosophical Magazine, 11, 592– 596.

Chandrasekhar, S., 1931, *The Maximum Mass of Ideal White Dwarfs*, ApJ, 74, 81–82.

Eddington, Arthur Stanley, 1934, *Relativistic Degeneracy*, Obs., 58, 37–39.

Eddington, Arthur Stanley, 1927, *Stars and Atoms*, Oxford University Press.

Fowler, R. H., 1926, *On Dense Matter*, MNRAS, 87, 114–122.

Landau, Lev D., 1932, *On the Theory of Stars*, Physikalische Zietschrift der Sowjetunion, 1, 285.

Miller, Arthur I., 2005, *Empire of the Stars*, Houghton Mifflin Co.

Shapiro, Stuart L. and Teukolsky, S. A., 1983, *Black Holes, White Dwarfs and Neutron Stars, the Physics of Compact Objects*, John Wiley & Sons.

Vibert, Douglas A., 1957, *The Life of Arthur Stanley Eddington*, Thomas Nelson & Sons, Ltd.

Wali, Kameshwar C.. 1991, *Chandra: A Biography of S. Chandrasekhar*, University of Chicago Press.

CHAPTER 9

Adams, Walter S., 1925, *The Relativity Displacement of Spectral Lines in the Companion of Sirius*, Proceedings of the National Academy of Sciences, 11, 382–387.

Adams, Walter S., 1925, *The relativity displacement of spectral lines in the companion of Sirius*, MNRAS, 48, 336–342.

Adams, W. S., 1926, *The Radial Velocity of the Companion of Sirius*, Obs., 49, 88.

Earman, J. and Glymour, C., 1980, *The Gravitational Redshift as a Test of General Relativity: History and Analysis*, Studies in History and Philosophy of Science, 11, 175–214.

Eddington, A. S., 1924, *On the Relation between the Masses and Luminosities of the Stars*, MNRAS, 84, 308–333.

Eddington, A. S., 1926, *The Internal Constitution of the Stars*, Cambridge University Press.

Eddington, A. S., 1941, *White Dwarfs Discovery, Observations, Surface Conditions*, in Novae and White Dwarfs Vol III, published by A. J. Shaler, Hermann & Co., Paris.

Einstein, A., 1911, *On the Influence of Gravitation on the Propagation of Light*, Ann. der Physik, 35, 898–908.

Greenstein, J. L.; Oke, J. B.; & Shipman, H. L., 1971, *Effective Temperature, Radius and Gravitational Redshift of Sirius B*, ApJ, 169, 563–566.

Greenstein, J. L.; Oke, J. B.; & Shipman, H. L., 1985, *On the Redshift of Sirius B*, Q. Jl. Astr. Soc., 26, 279–288.

Hetherington, Norris S., 1980, *Sirius B and the Gravitational Redshift: An Historical Review*, Q. Jl. Astr. Soc., 21, 246–252.

Joint eclipse meeting of the Royal Society and Royal Astronomical Society, 1919, Obs., 42, 389– 398.

Kodaira, K., 1967, *A Spectrum of Sirius B*, Publications of the Astronomical Society of Japan, 19, 172–179.

Kuiper, Gerard P., 1941, *White Dwarfs Discovery, Observations, Surface Conditions*, in Novae and White Dwarfs Vol III, published by A. J. Shaler, Hermann & Co., Paris.

Marshak, R. E., 1940, *The Internal Temperature of White Dwarf Stars*, ApJ, 92, 321–353.

Moore, J. H., 1928, *Recent Spectroscopic Observations of the Companion of Sirius*, PASP, 40, 229–233.

Pound, R. V. and Rebka, G. A. Jr., 1959, *Gravitational Red-Shift in Nuclear Resonance*, Physical Review Letters, 3, 439–441.

Pound, R. V. and Snider, J. L., 1964, *Effect of Gravity on Nuclear Resonance*, Phys. Rev. Letters, 13, 539–540.

Savaedoff, Malcolm P.; Van Horn, H. M.; Wesemael, F.; Auer, L. H.; Snow, T. P.; and York, D. G., 1976, *The Far-Ultraviolet Spectrum of Sirius B from Copernicus*, ApJ, 207, L45–L48.

Stromberg, G., 1926, *Note Concerning the Radial Velocity of the Companion of Sirius*, PASP, 38, 44.

Vessot, R. F. C., Levine, M. W., Mattison, E. M., Blomberg, E. L., Hoffman, T. E., Nysrom, G. U., Farrel, B. F., Decher, R., Eby, P. B., Baugher, C. R. *et al.*, 1980, *Test of Relativistic Gravitation with a Space-Borne Hydrogen Maser*, Physical Review Letters, 45, 2081–2084.

Vibert, Douglas A., 1957, *The Life of Arthur Stanley Eddington*, Thomas Nelson & Sons, Ltd.

Wesemael, F., 1985, *A Comment on Adam's Measurement of the Gravitational Redshift of Sirius B*, Q. Jl. Astr. Soc., 26, 273–278.

CHAPTER 10

Barker, Thomas, 1760, *Remarks on the mutations of the stars*, Phil. Trans., 51, 498–504.

Benest, Daniel and Duvent, J. L., 1995, *Is Sirius a triple star?*, A&A, 299, 621–628.

Bonnet-Bidaud, J. M.; Colas, F.; & Lecacheaux, J., 2000, *Search for Companions around Sirius*, A&A, 360, 991–996.

Brosch, Noah and Nevo, Isaac, 1978, *A Search for Nebulosity around Sirius*, Obs., 98, 136–137.

Bruhweiler, Frederick C.; Kondo, Yoji; and Sion, Edward M., 1986, *The historical record for Sirius evidence for a white-dwarf thermonuclear runaway?*, Nature, 324, 235–237.

Ceragioli, Roger C., 1993, *The Riddle of Red Sirius: An Anthropological Perspective, in Astronomies and Cultures*, eds. Clive, L. N. Ruggles and Nicholas J. Saunders, University of Colorado Press, 67–99.

Ceragioli, Roger C., 1995, *The Debate Concerning 'Red Sirius'*, JHA, 26, 187.

Ceragioli, Roger C., 1996, *Solving the Puzzle of 'Red Sirius'*, JHA, 27, 93.

Ceragioli, Roger C., 1996, *Behind the " Red Sirius" Myth*, Sky and Telescope, 83, 613–615.

Editor, 1899, *Remarks on Mr. Moultons' Paper in A.J. 461*, AJ, 20, 568.

Gry, C. and J. M. Bonnet-Bidaud, 1990, *Sirius and the colour enigma*, Nature, 347, 625.

Guinan, Edward, F. and Ribas, Ignasi, 2001, *The Best Brown Dwarf Yet? A Companion of the Hyades Eclipsing Binary V 471 Tauri*, ApJL, 546, 43–47.

Innes, R. T. A., 1929, *Sirius*, Obs., 656, 22–23.

Jiang xiao-yuan, 1993, *The colour of Sirius as recorded in ancient Chinese texts*, Chinese Astronomy and Astrophysics, 17/2, 223–228.

Johansson, Göran H. I., 1984, *Red Sirius*, Nature, 303, 568.

Kucher, Marc J., 2000, *A search for Exozodiacal Dust and Faint Companions near Sirius, Procyon, and Altair with the NICMOS Coronagraph*, PASP, 112, 827–832.

Lindenblad, Irving W., 1975, *On K. D. Rako's Photoelectric Measurements of Sirius B*, A&A, 41, 111–112.

McClusky, Stephan C., 1987, *The colour of Sirius in the sixth century*, Nature, 325, 87.

Moulton, Forrest R., 1899, *The Limits of Temporary Stability of Satellite Motion, with an Application to the Questions of the Existence of an Unseen Body in the Binary System F.70 Ophiuchi*, AJ, 20, 33–37.

Pennick, Nigel, 1984, *Red Sirius*, Nature, 312, 10.

Peterson, Charles J., 1992, *A Very Brief Biography and Popular Account of the Unparalleled T. J. J. See*, Griffith Observer, 7, 2–16.

Putnam, W. L., 1994, *The Explorers of Mars Hill*, Phoenix Publishing.

Rakos, K. D., 1974, *Photoelectric Measurements of Sirius B in the UBV and Strömgren System*, A&A, 34, 157–158.

Ridpath, Ian, 1984, *"Red" Sirius*, Obs., 108, 130.

Schroeder, Daniel J., Golimowski, D. A., Brukardt, R. A., Burrows, C. J., Caldwell, J. J., Fastie, W. G., Ford, H. C., Hesman, B., Kletskin, I., Krist, J. E. *et al.*, 2000, *A Search for Faint Companions to Nearby Stars Using the Wide Field Planetary Camera 2*, ApJ, 119, 906–922.

See, T. J. J., 1892, *History of the Color of Sirius*, Astronomy and Astro-Physics, 11, 269–274.

See, T. J. J., 1892, *History of the Color of Sirius*, Astronomy and Astro-Physics, 11, 372–386.

See, T. J. J., 1892, *Explanation of the Mystery of the Egyptian Pheonix*, Astronomy and Astro-Physics, 11, 457–461.

See, T. J. J., 1892, *Note on the History of the Color of Sirius*, Astronomy and Astro-Physics, 11, 550–552.

See, T. J. J., 1895, *Sirius in Ancient Times*, Popular Astronomy, 5, 193–198.

See, T. J. J., 1926, *Historical Researches Indicating a Change in the Color of Sirius between the Epochs of Ptolemy, 138, and of Al Sûfi, 980, A. D.*, AN, 229, 246–272.

Schlosser, Wolfhard and Bergmann, Werner, 1985, *An early-medieval account on the red colour of Sirius and its astrophysical implications*, Nature, 318, 45–46.

Schlosser, Wolfhard and Bergmann, Werner, 1987, *Reply to van Gent*, Nature, 25, 89.

Shaefer, Bradley E., 2000, *The Heliacal Rise of Sirius and Ancient Egyptian Chronology*, JHA, 31, 149–155.

Sheehan, William, 2002, *The Tragic Case of T. J. J. See*, Mercury, 35–39.

Sherrill, Thomas J., 1999, *A Career of Controversy: The Anomaly of T. J. J. See*, JHA, 30, 25–50.

Stephanides, Theodore, 1976, *Colour Changes of Sirius*, JBAA, 87, 93–94.

Tang, Tong B., 1986, *Star Colours*, Nature, 319, 532.

Tang, Tong B., 1991, *Did Sirius change color?*, Nature, 352, 25.

Torres, Carlos Alberto P. C. O., 1984, *Red Sirius*, Nature, 311, 8.

van Gent, Robert H., 1984, *Red Sirius*, Nature, 312, 302.

van Gent, Robert H., 1987, *Reply to Schlosser and Bergmann*, Nature, 325, 87–88.

van Gent, Robert H., 1989, *The Colour of Sirius*, Obs., 109, 23–24.

Violet, Ch., 1932, *Recherche des Perturbations dans le Système de Sirius*, Bull. Astron., Paris, 8, 51–64.

Webb, W. L., 1913, *Brief Biography and Popular Account of the Unparalleled Discoveries of T. J. J. See*, Thos. P. Nichols & Sons, Co.

Whittet, Douglas C. B., 1999, *A physical interpretation of the 'red Sirius' anomaly*, MNRAS, 310, 355–359.

Wright, R. H. and Cattley, R. E. D., 1983, *The wine dark sea*, Nature, 303, 568.

Zagar, F., 1931, *Il Terzo Corpo nel Sistema Sirio*, Pub. Roy. Obs. Astron. Padova, 23, 1049–1100.

CHAPTER 11

Aubert, Raphaël and Keller, Carl-A, 1994, *Vie et Mort De L'Ordre Du Temple Solaire*, Troisième partie L'Odre Initiatique du Temple Solaire, Editions de l'Aire/Jouvence.

Baize, Paul, 1931, *Le Compagnon de Sirius*, Bulletin de la Société Astronomique de France, 383–397.

Bedinni, Silvio A., 1972, *Life of Benjamin Banneker*, Scribner's & Sons.

Benest, D. and Duvent, J. L., 1995, *Is Sirius a triple star?*, A&A, 299, 621–628.

Brecher, Kenneth and Feirtag, M., 1979, *Astronomy of the Ancients*, chapter on *Sirius Enigmas*, MIT Press, 92–115,

Cerami, Charles, 2002, *Benjamin Banneker*, John Wiley & Sons, Inc.

Calame-Griaule, Geneviève, 1991, *On the Dogon Revisited*, Current Anthropology, 32, No. 5, 575–577.

Consolmagno, Guy, 2000, *Brother Astronomer: Adventures of a Vatican Scientist*, McGraw-Hill, 132.

Gatewood, George and Gatewood, Carolyn, 1978, *A Study of Sirius*, ApJ, 22, 191–197.

Griaule, Marcel and Dieterlen, Germaine, 1965, *Le Renard pâle*, Tome I, Fascicule 1, Institut d'Ethnologie, Musée de l'Homme, Palais de Chaillot, Place du Trocadero, Paris, 16e.

Griaule, Marcel and Dieterlen, Germaine, 1950, *Un Système Soudanais de Sirius*, Journal de la Société des Africainistes, XX, 2, 273–294.

Heusch, Luc de, 1991, *On Griaule on Trial*, Current Anthropology, 32, No. 4, 434–437.

Introvigne, Massimo, 1995, *Ordeal by fire: The tragedy of the Solar Temple*, Religion, 25, 267– 283.

Mayer, Jean François, 1998, *Apocalyptic Millennialism in the West: The Case of the Solar Temple*, Lecture at the University of Virginia.

Mayer, Jean François, 1999, *Our Terrestrial Journey Is Coming to an End: The Last Voyager of the Solar Temple*, Nova Religio, 2, 2, 172–196.

Palmer, Susan, 1996, *Purity and Danger in the Solar Temple*, Journal of Contemporary Religion, 11, No. 3, 303–318.

Sagan, Carl, 1979, *Broca's Brain*, Random House, 67–71.

Tabachnik, Michel, 1997, *Bouc Émissaire*, Michel Lafon, Paris.

Temple, Robert, 1998, *The Sirius Mysteries*, Destiny Books, Rochester, VT.

Van Beek, Walter E. A., 1991, *Dogon Restudied: A Filed Evaluation of the Work of Marcel Griaule*, Current Anthropology, 32, No. 2, 139–167.

Voltaire, *Micromégas*, translated by W. Fleming Dadalus/Hippocrene.

CHAPTER 12

Barstow, Martin A. and Holberg, Jay B., 2003, *Extreme Ultraviolet Astronomy*, Cambridge University Press.

Barstow, Martin A.; Bond, H. E.; Holberg, J. B.; Burleigh, M. R.; Hubeny, I.; and Koester, D., 2005, *Hubble Space Telescope spectroscopy of the Balmer lines in Sirius B*, MNRAS, 362, 1134–1142.

Gatewood, George and Gatewood, Carolyn, 1978, *A Study of Sirius*, ApJ, 22, 191–197.

Hanbury Brown, Robert and Twiss, Richard, Q., 1956, *A Test of a New Type of Stellar Interferometer on Sirius*, Nature, 178, 1046–1048.

Holberg, J. B.; Barstow, M. A.; Bruhweiler, F. C.; Cruise, A. M.; and Penny, A. J., 1998, *Sirius B: A New, More Accurate View*, ApJ, 497, 935–942.

Holberg, J. B.; Wesemael, F.; & Hubeny, I., 1984, *The Far-Ultraviolet Energy Distribution of Sirius B from Voyager 2*, ApJ, 280, 679–687.

Kervella, P.; Thévenin, F.; Morel, P.; Bordé, P.; and Di Falco, E., 2003, *The Interferometric diameter and the internal structure of Sirius A*, A&A, 408, 681–688.

Mewe, R., Heise, J., Gronenschild, E. H. B. M., Brinkman, A. C., Schrijver, J. and den Boggende, A. J. F., 1975, *Soft X-rays from Sirius*, Nature, 175, 256, 711–712.

Mewe, R., Heise, J., Gronenschild, E. H. B. M., Brinkman, A. C., Schrijver, J. and den Boggende, A. J. F., 1975, *Detection of X-Ray Emission from Stellar Coronae with ANS*, ApJ, 202, L67–L71.

Mozurkewich, D. *et al.*, 2003, *Angular Diameters of Stars from the Mark III Optical Interferometer*, AJ, 126, 2048–2059.

Perryman, M. A. C. *et al.*, 1997, *The Hipparcos Catalog*, A&A, 390, 611.

Shipman, Harry S., 1976, *Sirius B: Thermal Soft X-ray Source?*, ApJL, 206, L67–L69.

Van Altena, William F.; Truan-Lian Lee, T.; and Hoffleit, D. E., 1995, *General Catalog of Trigonometric Stellar Parallaxes*, Yale University.

van den Bos, Willem H., 1960, *The Orbit of Sirius, ADS 5423*, Journal des Observateurs, 43, 145–151.

CHAPTER 13

García-Sánchez *et al.*, 1999, *Stellar Encounters with the Oort Cloud Based on Hipparcos Data*, ApJ, 117, 1042–1055.

Hansen, Carl J.; Kawaler, Steve J.; and Trimble, Virginia, 2004, *Stellar Interiors, Physical Principles, Structure and Evolution*, Springer-Verlag.

Hertzsprung, Ejnar, 1909, *On New Members of the System of the Stars β, γ, δ, ε, ζ Ursae Majoris*, ApJ, 30,135–143.

Kervella, P.; Thévenin, F.; Morel, P.; Bordé, P.; and Di Falco, E., 2003, *The Interferometric diameter and the internal structure of Sirius A*, A&A, 408, 681–688.

Liebert, James; Young, P. A.; Arnett, D.; Holberg, J. B.; and Williams, K. A., 2005, *The Age and Progenitor Mass of Sirius B*, ApJL, 609, L69–L72.

Mukadam, A. S. *et al.*, 2003, *Constraining the Evolution of ZZ Ceti*, ApJ, 594, 961–970.

Rybicki, K. R. and Denis, C., 2001, *On the Final Destiny of the Earth and the Solar System*, Icarus, 151, 130–137.

Sackman, I.-J.; Boothroyd, A. I.; and Kraemer, 1993, *Our Sun, III: Present and future*, ApJ, 418, 457–468.

Soker, Noam, 1994, *The expected morphology of the Solar System planetary nebula*, PASP, 106, 59–62.

Ward, Peter D. and Brownlee, Donald, 2003, *The Life and Death of the Planet Earth*, Times Books, Henry Holt & Co.

Index

Printing: Mercedes-Druck, Berlin
Binding: Stein + Lehmann, Berlin